CHEMICAL THERMODYNAMICS

Revision and Worked Examples

CHEMICAL THERMODYNAMICS

Revision and Worked Examples

H. P. STADLER
Department of Chemistry
University of Newcastle upon Tyne

British Library Cataloguing in Publication Data
Stadler, H. P.
Chemical thermodynamics: revision and worked examples.
1. Chemical reactions. Thermodynamics
I. Title II. Royal Society of Chemistry
541.3′69

ISBN 0-85186-273-X

Published by The Royal Society of Chemistry
Thomas Graham House, Cambridge CB4 4WF

Typeset by KEYTEC, Bridport, Dorset
Printed by Whitstable Litho, Whitstable, Kent

Preface

Having had many years' experience of teaching thermodynamics to chemists, metallurgists, and chemical engineers, my philosophy has always been that, to be useful, thermodynamics must be applied in examples and that these will also further the understanding of the rather forbidding symbolism used. This book therefore sets out to teach thermodynamics through its applications. The theory of the subject is only presented in short revision form, but covers the syllabus required by the Institution of Chemical Engineers. This is supplemented by copious worked examples, with special attention given to areas which students find difficult, and by the use of the simplest methods and useful approximations. The necessary mathematics is given in an appendix, as are conversion factors from non-SI to SI units to enable readers to utilize earlier data in the literature. Example calculations for that are included.

H. P. Stadler
May 1989

Introduction

I have arranged the chapters in ascending order of complexity rather than according to subject matter only. All ideal systems are therefore discussed in the early chapters 1 to 4 before dealing with non-ideal and open systems in chapters 5 to 9. I have also tried to grade the examples which follow the brief exposition of the theory in each chapter from simple to more complex.

The examples were selected to illustrate the theory without being repetitive and are given at the end of each section of revision, followed by a block of answers. It hardly requires stressing that these answers are designed for readers who have first attempted the examples. Approximations are encouraged, where the required accuracy permits, but the order of magnitude of the error must be taken into account so as not to imply excessive accuracy in the answers.

Appendices deal with the required mathematics and with constants, units, and conversion factors.

IUPAC NOMENCLATURE

I have tried, wherever possible, to use the new IUPAC symbols and notation,[1,2] although it will be some time before all undergraduate textbooks adopt all of them. In the interest of simplicity I have, however, adopted the symbol γ for activity coefficients, both for solvents and solutes (see Chapter 7), and even for fugacity coefficients. Where necessary, subscripts are used to differentiate between different reference states. The letter f, suggested by IUPAC for activity coefficients of mixtures and also for fugacities, has been reserved for fugacities only (to avoid confusion, and in accord with many current textbooks).

A list of symbols is given in Appendix C together with any IUPAC alternatives and page references.

CONVENTIONS

I have followed the IUPAC recommendations to represent quantities as pure numbers × units (or quantity/units = pure number) and tried to ring the changes to familiarize the reader with the various conventions. To save overburdening the answers to examples with units, I shall 'carry forward' the units to the end of an equation, so that the $\mathrm{J\,mol^{-1}}$ in $31\,400 + 43.5 \times 37 = 33\,009\,\mathrm{J\,mol^{-1}}$ is intended equally for 31 400 and 43.5 × 37 as if the units had to be multiplied out throughout the bracketed equation $(31\,400 + 43.5 \times 37 = 33\,009)\,\mathrm{J\,mol^{-1}}$. The multiplication of a string of units takes precedence over other multiplications or divisions, so that $\Delta H/\mathrm{J\,mol^{-1}} = \Delta H/(\mathrm{J\,mol^{-1}})$.

The multiplication rule or algebra of quantities applies equally to units of kg or to the quantity 'amount of substance' with units of mol. Quantities per unit mass are called 'specific' (represented by lower case symbols), those relating to 1 mol 'molar', where the mole should be specified by an equation or formula. I have relied on the good sense of the reader to supply this definition in context. In particular, I have usually omitted the subscript m for molar, though I have retained the distinction between, say, the volume V and the partial molar volume V' $(\bar{V})$ of a substance in a mixture.

To simplify, I have omitted the all-pervasive standard sign $(^{\ominus})$ in expressions like $\Delta_f H$ and $\Delta_c H$, for which the standard pressure, $P^{\ominus}$, is specified in the definition of these quantities. $P^{\ominus}$ is fundamental to thermodynamic calculations since most data compilations are measured at $P^{\ominus}$. Since 1982, $P^{\ominus} = 10^5$ Pa = 1 bar. The old standard of 1 atm (= 10 125 Pa) is, however, still frequently used or implied in the pre-1982 literature. The value of $P^{\ominus}$ in the examples is therefore chosen appropriate to the data. The small difference between old and new $P^{\ominus}$ can normally be neglected for changes *not* involving gases.

Great care is crucial with any units, and while SI base units are consistent, errors of several magnitudes may be introduced if these base units are confused with other units *e.g.* Pa with $P^{\ominus}$, or $\mathrm{cm^3}$ with $\mathrm{m^3}$. Since dimensionless ratios are an easy way of avoiding units, this leads to the introduction of other 'standards' such as $v^{\ominus}$ (usually $1\,\mathrm{cm^3\,g^{-1}}$), $m^{\ominus}$ $(= 1\,\mathrm{mol\,kg^{-1}})$, as well as $P^{\ominus}$ as accepted devices to get pure numbers that can be readily used in calculations; but they do unfortunately complicate formulae. I shall therefore introduce the superscript solidus symbol $(^{/})$

to define the dimensionless ratios $T' = T/\mathrm{K}$, $m' = m/m^\ominus$, and $P' = p/P^\ominus$.

As a further simplification I shall neglect the 0.15 in °C into K conversions, such as the important reference temperature of 25 °C, to which most recent data compilations refer, and omit units from subscripts and superscripts if no confusion is likely to arise, *e.g.* in ΔH_{298} or integral limits.

Throughout my teaching of thermodynamics I have had to compromise between being pedantically precise, using multiple subscripts or superscripts and long bracketed expressions which obstruct the overall view, and misleadingly over-simplifying. I believe that the extent of this compromise must depend on the stage reached by the reader, and I hope that the reader who has followed the argument will be able to provide the correct conditions himself. It would be impossible to satisfy all, but I hope to be helping most, at least some of the time.

My thanks are due to my students over many years, who clamoured for a book of revision and worked examples, to Professor D. H. Whiffen for valuable help with IUPAC conventions, and to Dr J. H. Carpenter and other colleagues for discussions on the teaching of thermodynamics. Professor W. Ledermann has very kindly offered expert guidance on Appendix A, and I am particularly grateful to Professor A. K. Covington for his advice and encouragement, and for providing facilities. I am greatly in debt to Mr W. H. Beck for his most thorough examination of the text and all the examples. Without his detailed comments, many errors and ambiguities might have remained. Finally, I must acknowledge the help and advice given by Dr R. H. Andrews and the editorial staff at the Royal Society of Chemistry.

Contents

Chapter 1

The First Law of Thermodynamics

1.1 INTERNAL ENERGY

The First Law is really a statement of the Principle of Conservation of Energy ('Energy can neither be created, nor destroyed') with special reference to heat energy and work done. It says that heat energy can either be converted into work, or be stored. In mathematical terms† it is

$$\delta q = \delta w + \mathrm{d}U \tag{1.1}$$

In this equation δq and δw are infinitesimal quantities of, respectively, the heat adsorbed and the work done by‡ a system (such as a given quantity of gas, liquid, solid, or any mixture of them). The remaining item, $\mathrm{d}U$, balances the equation, as required by the Principle of Conservation of Energy.

U is known as the *internal energy* and can be regarded as the energy stored by the system. This stored energy must itself be subject to the conservation principle, and any changes in it are therefore fixed by the initial and final states of the system (as described by its *state variables*, for example T, P, amount, *etc.*). Thus, if a certain amount of heat is supplied to a gas, say, some

†The requisite mathematical symbolism is discussed in Appendix A.

‡In many books, the first law expression is $\mathrm{d}U = \delta q + \delta w$ with $\delta w = -P\mathrm{d}V$ in equation 1.3. After this substitution for δw, equation 1.4 and later equations remain unchanged. The description of this δw as 'work done *on* the system' is, however, only correct for $P_{ext} = P_{sys}$, *i.e.* at equilibrium; it could mislead, which prompted my alternative choice of sign convention for δw.

of that heat may be converted into work of expansion, the remainder being stored in the gas itself as extra (kinetic and other) energy of the gas molecules and their interactions – the gas will become hotter and possibly change its pressure. If the system undergoing the change is a boiling liquid, some of the heat will go into the conversion from liquid into gas, without any change of temperature. In each case, the increase in stored energy is characteristic of the change of state of the system (defined by its T, V, P, state of aggregation, *etc.*), and this makes the internal energy U a *state function*; hence the use of the symbol d (as opposed to δ), as outlined in Appendix A.

The most convenient state variables for U are T and V [often written as $U = U(T, V)$], so that equation A.5 (see Appendix A, p.120) becomes

$$dU = (\partial U/\partial T)_V dT + (\partial U/\partial V)_T dV \tag{1.2}$$

The work term δw may include electrical, magnetic, and gravitational work, but the most important term is usually work of expansion against an external pressure, P. This exerts a force PA on a piston of area A (see Figure 1.1), so that

$$\delta w = PA\,dx = P\,dV \tag{1.3}$$

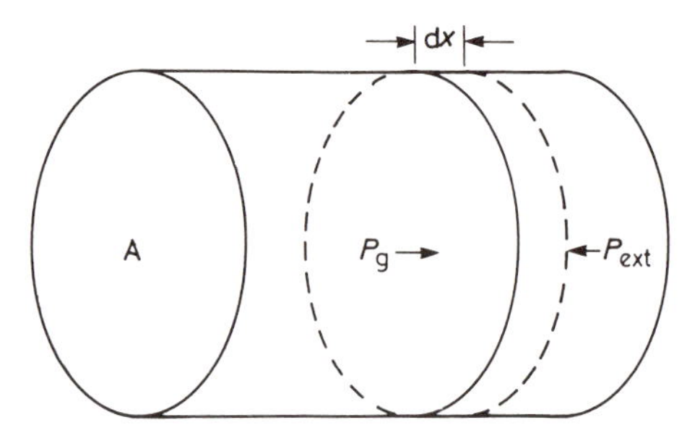

Figure 1.1 *Work of expansion. The force on the piston is PA, therefore the work done (force × distance) is $\delta w = PA\,dx = P\,dV$ ($dV = A\,dx$ is the volume swept out by the piston). The opposing pressure is P_{ext}; P_g is irrelevant, except that it has to be larger than P_{ext} for an expansion to occur*

1.2 ENTHALPY

For work of expansion only, therefore, the First Law expression equation 1.1 becomes

$$\delta Q = \mathrm{d}U + P\mathrm{d}V \tag{1.4}$$

New state functions can be constructed from suitable combinations of state functions and state variables, since both depend only on the state of the system. It is convenient to introduce a state function, the *enthalpy*, H, defined by the equation

$$H = U + PV \tag{1.5}$$

with P and V state variables of the system. This gives (by equation 1.4) an exact differential:

$$\mathrm{d}H = \mathrm{d}U + \mathrm{d}(PV) = \mathrm{d}U = P\mathrm{d}V + V\mathrm{d}P \tag{1.6}$$

or

$$\mathrm{d}H - V\mathrm{d}P = \mathrm{d}U + P\mathrm{d}V = \delta q \tag{1.7}$$

providing an alternative First Law expression:

$$\delta q = \mathrm{d}H - V\mathrm{d}P \tag{1.8}$$

1.3 MOLAR HEAT CAPACITIES

For constant volume ($\mathrm{d}V = 0$ in equation 1.4) $\delta q_V = \mathrm{d}U$, and similarly at constant pressure $\delta q_P = \mathrm{d}H$ (equation 1.8), so that under these conditions the heat absorbed becomes equal to a change in a state function. It also follows that the heat required to raise the temperature of one mole of substance by 1 K, the molar heat capacity at constant volume, is

$$C_V = (\partial q/\partial T)_V = (\partial U/\partial T)_V \tag{1.9}$$

and at constant pressure

$$C_P = (\partial q/\partial T)_P = (\partial H/\partial T)_P \tag{1.10}$$

where U and H must also refer to one mole of substance.

1.4 HESS'S LAW

The advantage of the state functions X, such as the internal energy U or the enthalpy H, is that ΔX is, by definition (see Appendix A), independent of the path, so that once calculated for one set of conditions they remain valid irrespective of the path or any intermediate stages on the way. This, indeed, is the justification for Hess's Law, which holds provided the heat of reaction is either at constant volume or at constant pressure, when the heat absorbed becomes ΔU or ΔH, respectively, and thus the change in a state function.

Since most reactions are performed at constant pressure (viz. atmospheric pressure), ΔH and C_P are the most used functions, but conversion is simple. Where no gases are involved the difference between ΔU and ΔH can usually be neglected. For any gases at low pressures ideal gas behaviour ($PV = nRT$) may be assumed, so that

$$\Delta H = \Delta U + \Delta(PV) = \Delta U + \Delta n(RT) \quad \text{(at constant } T\text{)} \tag{1.11}$$

where Δn is the change in the amount of gas (moles) in the reaction.

All that is required to calculate ΔH for any reaction is a set of tabulated reference reactions for any compound. The most readily accessible data for organic substances are for their combustion reactions, and their enthalpies, $\Delta_c H$, are widely used to calculate $\Delta_f H$, the ΔH for the formation reaction from the elements in their reference states, *i.e.* the stable form of the element (*e.g.* graphite for carbon) at the temperature considered and at the *standard pressure*, $P^\ominus$. This used to be 1 atm (= 101.325 kPa), but is now defined as 1 bar (= 10^5 Pa) (see Appendix B). There are extensive tabulations of the enthalpy of formation of organic and inorganic compounds at 25 °C (298.15 K), usually written $\Delta_f H_{298}$. (Since $\Delta_f H$ always refers to the standard state, the superscript $^\ominus$ in $\Delta_f H^\ominus$ or $\Delta_c H^\ominus$ is redundant and will be omitted.)

Figure 1.2 shows how the use of $\Delta_c H$ and $\Delta_f H$ leads to the general results

$$\Delta H^\ominus = \Sigma\, \Delta_f H \tag{1.12a}$$

and

$$\Delta H^\ominus = -\Sigma\, \Delta_c H \tag{1.12b}$$

with both summations over products minus reactants (final − initial states) (see Appendix A and examples).

Since both reactants and products of the reaction must give the same combustion products in equal amounts, $\Delta_c H$ for the reactants must be the same as ΔH for the reaction plus $\Delta_c H$ for the products. Note also that $\Delta_f H$ and $\Delta_c H$ are normally given for one

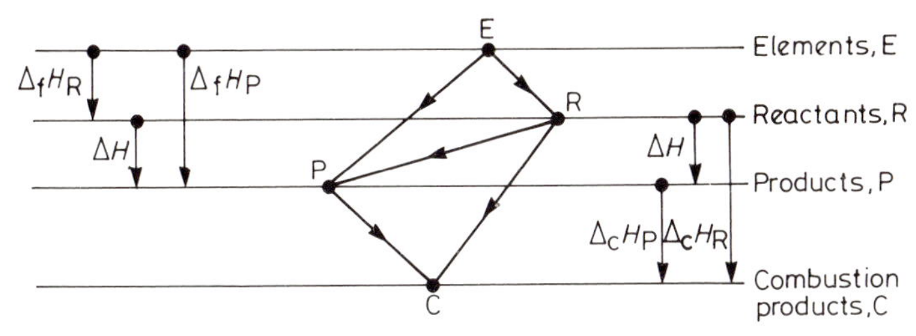

Figure 1.2 *Diagram for state functions. State functions have fixed levels for any given state (in this instance, H is the example used). Only the 'level differences' matter. Triangles indicate the alternative paths to be followed and help to get the signs right* (e.g. $ER + RP = EP$, $\therefore RP = EP - ER$; *or* $RP + PC = RC$, $\therefore RP = RC - RP$)

mole of substance, and have to be multiplied by their *stoichiometric coefficients*, ν, if several moles are involved in a reaction.

All the arrows in Figure 1.2 go downwards to a lower level, and their ΔH values will all be negative. Reactions going in the opposite direction will be equal in value but opposite in sign, *i.e.* positive. Such 'uphill' reactions will absorb heat and are termed *endothermic*, whereas 'downhill' reactions give off heat and are *exothermic* (see equation 1.8). The same applies to ΔU changes at constant volume.

1.5 KIRCHHOFF'S EQUATION

The same principle (that change in a state function is independent of the path) can also be used to extend the data from one particular temperature to other temperatures, but care must be taken to include all the changes in the state functions involved. In particular, any phase transitions will, of course, change the state of the system (see Chapter 4).

$$\begin{array}{ccccc} \text{Reactants} & \xrightarrow{\Delta X_1} & \text{Products} & & T_1 \\ \downarrow & & \downarrow & & \\ \text{Reactants} & \xrightarrow{\Delta X_2} & \text{Products} & & T_2 \end{array}$$

By equating the changes obtained by two different routes (see equation A.8, p.120):

$$\Delta X_1 + \int_1^2 (\partial X/\partial T)_{pr}\,dT = \int_1^2 (\partial X/\partial T)_{re}\,dT + \Delta X_2 \quad (1.13)$$

or

$$\Delta X_2 = \Delta X_1 + \int_1^2 \partial\Delta X/\partial T\,dT \text{ (see equation A.3)}$$

If X is the enthalpy, $\partial X/\partial T$ becomes C_P, $\partial\Delta X/\partial T$ becomes $\Delta C_P = (C_P)_{pr} - (C_P)_{re}$, and equation 1.13 becomes

$$\Delta H_2 = \Delta H_1 + \int_1^2 \partial\Delta H/\partial T\delta T = \Delta H_1 + \int_1^2 \Delta C_P\,dT \quad (1.14)$$

This is sometimes called Kirchhoff's equation, and it applies to temperature changes without any phase changes, although it can be extended to include them:

$$\Delta H_2 = \Delta H_1 + \int_1^2 \partial\Delta H/\partial T\,dT + \Sigma\,\Delta_{tr}H \quad (1.15)$$

with the summation over products minus reactants.

1.6 IDEAL GAS EXPANSION

Ideal gases have played an important role both in the development of thermodynamic theory and as model substances for real gases and vapours. As an example, let us return to work of expansion against an external pressure, which we will denote by P_{ext} so as not to confuse it with the pressure exerted by the gas inside the cylinder, P_g (see Figure 1.1). For any gas at constant temperature, P_g would follow a curve called an *isotherm*. For an ideal gas $P_g = nRT/V$ and its isotherm is shown in Figure 1.3. Expansion can only take place with $P_g > P_{ext}$. Since P_{ext} can have any value from 0 to the limiting case P_g, the work done by the gas will likewise vary from 0 to $P_g\,dV$.

In the latter case the external pressure would have to keep pace with the falling gas pressure P_g, and the piston would go through a series of equilibrium states, so that its direction could be reversed by the slightest increase of P_{ext}. The work done in such a reversible isothermal expansion would be the area under the curve, and would clearly be the maximum possible work between the two states V_1 and V_2. It is the change in a state function, the 'maximum work function' or 'Helmholtz energy'. It should also be pointed out that the same expression holds for both expansion and compression, which only differ in the relative magnitudes of

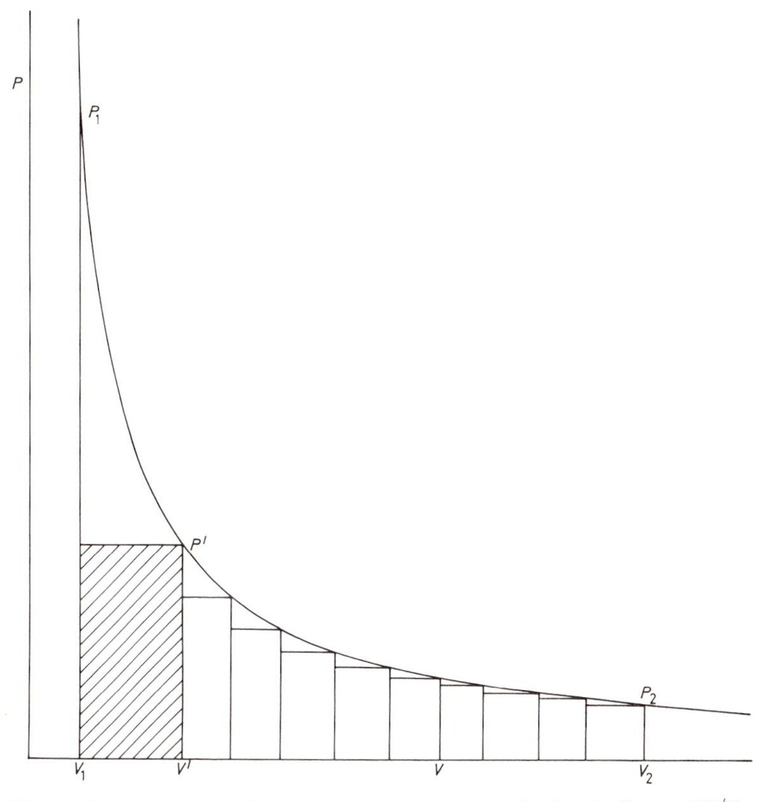

Figure 1.3 *Ideal gas isotherm. At constant temperature (isotherm), $P_g = nRT/V$ follows the curve shown. If the gas starts at P_1 and expands against a fixed external pressure, P'_{ext}, the gas would expand to V' with $w = (V' - V_1)P'_{ext}$, the shaded area. To obtain further work the external pressure would have to be reduced to be on or below the P_g line. If it follows the line to P_2 through a series of equilibrium positions ($P_{ext} = P_g$), the work done would then be a maximum,* i.e. *the whole area under the curve*

V_1 and V_2.

In an ideal gas there are no intermolecular interactions, so that the internal energy depends only on the temperature, not on the volume or pressure. (See also equation 2.16 and note that $dH = dU + R\,dT$ for an ideal gas.) Therefore

$$(\partial U/\partial V)_T = (\partial H/\partial P)_T = 0 \quad (1.16)$$

In an isothermal reversible expansion of an ideal gas, therefore, both right-hand terms in equation 1.2 are 0 so that $\mathrm{d}U = 0$. Thus by equation 1.1 $\delta q = \delta w$, or

$$q = w = \int_1^2 (nRT/V)\,\mathrm{d}V = nRT \ln (V_2/V_1) \tag{1.17}$$

In practice, the expansion would have to be carried out exceedingly slowly to ensure thermal equilibrium with the environment. At the other extreme, if no heat exchange is allowed (*e.g.* if the expansion is very rapid compared with the rate of heat transfer, as in the propagation of sound waves), the expansion is termed *adiabatic*, and $\delta q = 0$. By equation 1.1, any work done by the gas would then be at the expense of the internal energy, *i.e.* $\delta w = -\mathrm{d}U$. Since for an ideal gas U is independent of the volume (see equation 1.16), and C_V can be shown to be a constant by simple kinetic theory:

$$\mathrm{d}U = C_V\,\mathrm{d}T \tag{1.18a}$$

and

$$\mathrm{d}H = C_P\,\mathrm{d}T \tag{1.18b}$$

$$\therefore\ w = -\int_1^2 (\partial U/\partial T)_V\,\mathrm{d}T = -\int_1^2 C_V\,\mathrm{d}T = -C_V(T_2 - T_1) \tag{1.19}$$

It also follows from equation 1.7 that for an adiabatic expansion of an ideal gas

$$V\,\mathrm{d}P = C_P\,\mathrm{d}T \tag{1.20}$$

and

$$P\,\mathrm{d}V = -C_V\,\mathrm{d}T \tag{1.21}$$

Dividing and putting $\mathrm{d}\ln x$ for $\mathrm{d}x/x$ gives

$$-\mathrm{d}\ln P/\mathrm{d}\ln V = C_P/C_V \tag{1.22}$$

Integrating and taking antilogs gives

$$P^{-1} = V^{C_P/C_V} + I' \tag{1.23a}$$

or

$$PV^{C_P/C_V} = I' \tag{1.23b}$$

where I' is the constant of integration.

EXAMPLES

(Data in the literature are frequently in non-SI units; conversion factors in Appendix B may be used to present answers in SI units.)

1.1 Calculate (to two-figure accuracy) the external work done by the system during the following changes at 1 atm, per mole of water.
(a) The melting of ice at 0 °C.
(b) The heating of water from 0 to 100 °C.
(c) The boiling of water at 100 °C.
(d) As for (c), but assuming steam to behave as an ideal gas.
The relative densities, d, (taking water at 4 °C as $d = 1$) are:
at 0 °C : ice, 0.917; water, 0.9999;
at 100 °C : water, 0.9584; steam, 0.60×10^{-3}.

1.2 What is meant by a 'state function'?
If X is a state function and $dX = \delta q_{rev}/T$, where δq_{rev} stands for reversibly absorbed heat, calculate ΔX for the isothermal expansion of one mole of an ideal gas from 2 to 1 bar at 298 K.

1.3 For each of the two paths (a) and (b) below, calculate the work done, the change in internal energy, and the heat absorbed in expanding one mole of an ideal gas from an initial volume of 10 litres and 0 °C to a final volume of 20 litres and 100 °C (taking $C_v = 20\ \mathrm{J\ K^{-1}\ mol^{-1}}$).
(a) A reversible isothermal expansion at 0 °C from 10 to 20 litres followed by heating at constant volume to 100 °C.
(b) Heating 10 litres at constant volume to 100 °C followed by a reversible isothermal expansion at 100 °C to 20 litres.
Compare the results, and comment.

1.4 $2C_2H_2(g) + 5O_2 = 4CO_2 + 2H_2O$, $\Delta H_{298} = -2510\ \mathrm{kJ\ mol^{-1}}$
$2C_6H_6(l) + 15O_2 = 12CO_2 + 6H_2O$,
$\Delta H_{298} = -6340\ \mathrm{kJ\ mol^{-1}}$
Calculate ΔH and ΔU for the reaction $3C_2H_2(g) = C_6H_6(l)$.

1.5 The heat evolved during the combustion of naphthalene, $C_{10}H_8$, in a bomb calorimeter (ΔU or ΔH?) was found to be 1234.6 kcal mol^{-1} at 15 °C. At the same temperature the values for $\Delta_c H$ of C and H_2 are, respectively, −94.0 and −68.4 kcal mol^{-1}. Calculate $\Delta_f H$ for naphthalene in kJ mol^{-1}. [1 cal = 4.184 J: see Appendix B.]

1.6 The mean molar heat capacities of $CHCl_3$ between 25 and 62 °C are 118.0 J K^{-1} mol^{-1} for the liquid and 74.5 J K^{-1} mol^{-1} for the vapour. The enthalpy of vaporization at 62 °C is 31.4 kJ mol^{-1}. Estimate the value at 25 °C.

1.7 From the heats of combustion for 300 K and the coefficients given below of the heat capacity equation

$$C_P/\text{cal K}^{-1}\text{ mol}^{-1} = a + bT' + c(T')^{-2} \text{ (with } T' = T/\text{K)}$$

calculate ΔH in kJ mol^{-1} for the water-gas reaction at 1300 K:

$$C + H_2O\,(g) = CO + H_2$$

State whether the reaction is endothermic or exothermic.

	$\Delta_c H$/kcal mol^{-1}	a	$b \times 10^3$	$c \times 10^{-5}$
C	−94.40	4.04	1.20	−0.17
CO	−67.95	6.79	0.98	−0.11
H_2	−57.80	6.52	0.78	0.12
H_2O	–	7.17	2.56	0.08

1.8 Calculate the heat of formation of 1 mole of dilute hydrochloric acid from the following calorimetric measurements (at the same T).
[*N.B.* 'aq' stands for aqueous solution. Heats of solution may be assumed to be additive to a sufficient approximation.]

Reaction	ΔH/kJ mol^{-1}
$S + O_2 = SO_2$	−296.8
$SO_2 + aq = SO_2(aq)$	−32.2
$SO_2(aq) + Cl_2 + 2H_2O = H_2SO_4(aq) + 2HCl(aq)$	−309.2
$S + 2O_2 + H_2 + aq = H_2SO_4(aq)$	−882.4
$H_2 + \frac{1}{2}O_2 = H_2O(l)$	−285.8

1.9 Estimate (to ±20 K) the adiabatic flame temperature of methane in air. (Assume that 5 mol air contain 1 mol O_2 with 4 mol N_2 and that all the gases start at an initial temperature of 298 K. The adiabatic flame temperature is the temperature the combustion products would attain without any heat loss to their surroundings.) For CH_4, $\Delta_c H = -890$ kJ mol^{-1}.

Using the following heat capacity equations for $C_P/\text{J K}^{-1}\,\text{mol}^{-1}$:

for CO_2, $32.2 + 22 \times 10^{-3}T' - 3.5 \times 10^{-6}(T')^2$

for H_2O, $35.0 + 0.7 \times 10^{-3}T' + 5.0 \times 10^{-6}(T')^2$

for N_2, $28.3 + 2.6 \times 10^{-3}T' + 0.5 \times 10^{-6}(T')^2$

1.10 Using the hydrostatic pressure equation, $\mathrm{d}P = g\rho\,\mathrm{d}z$, for the increase in fluid pressure with increasing column height $\mathrm{d}z$, and assuming dry air masses (with $c_P = 1.0\ \text{kJ K}^{-1}\ \text{kg}^{-1}$) to move up and down the atmosphere without heat exchange, calculate the 'dry adiabatic lapse rate', $\mathrm{d}T/\mathrm{d}z$, in free air.

ANSWERS

A1.1 The molar volume $V_m/\text{cm}^3\,\text{mol}^{-1} = 18/d = 19.6$, 18.00, 18.78, and 30 000 for ice, water at 0 °C, water at 100 °C, and steam, respectively. Taking $P = 101$ kPa as sufficiently accurate:

$$P(V_2 - V_1) = 101 \times 10^3(\Delta V/\text{cm}^3) \times 10^{-6}\ \text{J}$$

giving the results (a) **−0.16 J**, (b) **+0.08 J**, and (c) **+3.0 kJ** for 1 mole of water each. Assuming steam to behave as an ideal gas, $P\Delta V \approx RT = 8.314 \times 373 =$ **3.1 kJ**.

Results (a) and (b) show that work involving condensed phases (solids and liquids) only is negligible compared with that involving gases.

A1.2 A state function is fully determined by its state variables (see Appendix A). It follows that changes in state functions only depend on the initial and final state, and are independent of the path. If X is a state function, ΔX by any path must be equal to ΔX by a reversible path, *e.g.* $\int \delta q_{\text{rev}}/T$. By equation 1.17,

$$\delta q_{\text{rev}}/T = nRT/V\,\mathrm{d}V$$

$$\therefore\ \delta q_{\text{rev}}/T = nR\,\mathrm{d}\ln V$$

and

$$\Delta X = R\ln(V_2/V_1)$$

but for an ideal gas

$$V_2/V_1 = P_1/P_2 = 2$$

$$\therefore \Delta X = R \ln 2 = \mathbf{5.76\ J\ K^{-1}\ mol^{-1}}.$$

A1.3 For isothermal reversible expansions use equation 1.17 for q and w, and note that $\Delta U = 0$.

For heating at constant volume use the integrated form of equation 1.9 to obtain q and ΔU, but note that $\delta w = P\,\mathrm{d}V = 0$. For 1 mol:

(a) expansion step: $q = w = 273R \ln 2 = 1573\ \mathrm{J}, \Delta U = 0$

heating step: $q = \Delta U = 100C_V = 2000\ \mathrm{J}, \quad w = 0$

$\therefore$ overall: $w = \mathbf{1573\ J}, q = \mathbf{3573\ J}, \Delta U = \mathbf{2000\ J}$

(b) heating step: $q = \Delta U = 100C_V = 2000\ \mathrm{J}, w = 0$

expansion step: $q = w = 373R \ln 2 = 2150\ \mathrm{J}, \quad \Delta U = 0$

$\therefore$ overall: $w = \mathbf{2150\ J}, q = \mathbf{4150\ J}, \Delta U = \mathbf{2000\ J}$

While q and w differ according to the path taken, their difference, ΔU, is independent of the path, as would be expected since U is a state function.

A1.4 Using equation 1.12, and noting that the given data are for 2 moles:

$$\Delta H = -[\Delta_c H(C_6H_6) - 3\Delta_c H(C_2H_2)] = 3170 - 3765$$

$$= \mathbf{-595\ kJ\ mol^{-1}}$$

By equation 1.11

$$\Delta U = \Delta H - \Delta nRT$$

with $\Delta n = -3$ moles of gas

$$\therefore \Delta U = -595 + 3 \times 8.314 \times 298 \times 10^{-3}$$

$$= \mathbf{-588\ kJ\ mol^{-1}}$$

Note that the temperature only refers to the starting and end points of the reactions and that any state function is independent of the temperature (or any other) variations between these points.

A1.5 To ensure correct signs, a diagram clearly showing the direction of reactions will be found most useful. The single unknown change in a state function can then be found by equating the results of two alternative paths.

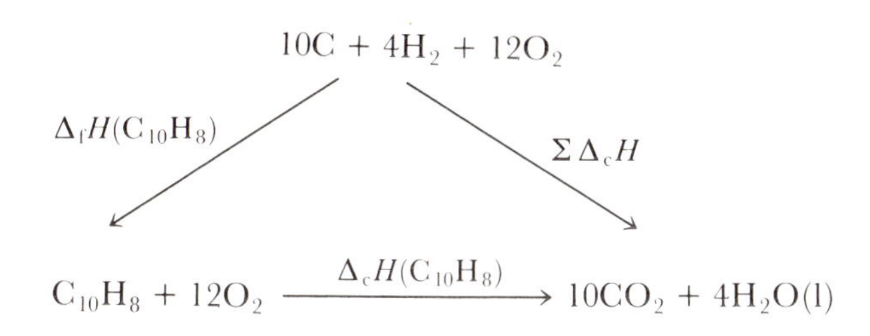

Following the arrows in this example, $\Delta_f H(C_{10}H_8) + \Delta_c H(C_{10}H_8)$ must equal $\Sigma \Delta_c H$ for the elements. (Equation 1.12 would give the same result.)

$$\therefore \ \Delta_f H(C_{10}H_8) = 10\Delta_c H(C) + 4\Delta_c H(H_2) - \Delta_c H(C_{10}H_8)$$

Note, however, that a bomb calorimeter works at constant volume, so that the heat envolved is $-\Delta U$ and has to be converted into ΔH by equation 1.11 with $\Delta n = -2$ (CO_2 and O_2 being the only gases involved), so that

$$\Delta_c H(C_{10}H_8) = \Delta_c U(C_{10}H_8) - 2RT = -1235.8 \text{ kcal mol}^{-1}$$

$$(R \approx 2 \text{ cal K}^{-1} \text{ mol}^{-1}, \text{ making } 2RT = 1.2 \text{ kcal mol}^{-1})$$

$$\therefore \ \Delta_f H(C_{10}H_8) = -940 - 273.6 + 1235.8$$

$$= 22.2 \text{ kcal mol}^{-1} = \mathbf{92.9 \text{ kJ mol}^{-1}}$$

A1.6 Thermodynamic equations apply equally to chemical or physical changes. Using equation 1.14 for $CHCl_3$ liquid to vapour

$$\Delta H_{298} = \Delta H_{335} + \int_{335}^{298} \Delta C_P \, dT$$

with

$$\Delta C_P = -43.5 \text{ J K}^{-1} \text{ mol}^{-1}, \ T_2 - T_1 = -37 \text{ K}$$

$$\therefore \ \Delta H_{298} = 31\,400 + 43.5 \times 37 = 33\,009 \text{ J mol}^{-1}$$

$$= \mathbf{33.0 \text{ kJ mol}^{-1}} \ (\textit{not } 33.009)$$

A1.7 It avoids unnecessary effort and occasion for mistakes if (a) conversion of units is carried out at the end only, and (b) summations are carried out before integration (see Appendix A).

$$\Delta H = -(-67.95 - 57.80 + 94.40) = 31.35 \text{ kcal mol}^{-1}$$
$$= 131.2 \text{ kJ mol}^{-1}$$

and the coefficients of the ΔC_P equation are 2.10, -2.00×10^{-3}, and 10^4

$$\therefore \Delta H_{1300} = \Delta H_{300} + 2100 - 1600 + 25 \text{ cal mol}^{-1}$$
$$\therefore \Delta H_{1300} = 31\,350 + 525 = 31\,880 \text{ cal mol}^{-1}$$
$$= \mathbf{133.4\ kJ\ mol^{-1}}$$

This means there is heat absorbed and the reaction is *endothermic*. (Note the small change in ΔH over a wide range of T.)

A1.8 Number the given chemical equations 1—5 and the required equation becomes equation 6:

$$\tfrac{1}{2}H_2 + \tfrac{1}{2}Cl_2 + aq = HCl(aq) \ \Delta H_6 = ?$$

Judicious manipulation – note that considering the compounds will automatically make the elements and aq come right – or the formal application of equation 1.14 to equation 3 and solving for ΔH_6 will then give:

$$\text{eq.3} - \text{eq.4} + 2 \times \text{eq.5} + \text{eq.2} + \text{eq.1} = 2 \times \text{eq.6}$$

$$\therefore 2\Delta_f H[HCl(aq)] = \Delta H_3 - \Delta H_4 + 2\Delta H_5 + \Delta H_2 + \Delta H_1$$
$$= -327.4 \text{ kJ mol}^{-1}$$

or

$$\Delta_f H[HCl(aq)] = \mathbf{-163.7\ kJ\ mol^{-1}}$$

A1.9 Formally, the process can be regarded as adiabatic overall ($\Delta H = 0$) with a stage 1 as the oxidation reaction at 298 K and stage 2 the heating of the combustion products (plus the inert nitrogen in the air) to the flame temperature.

The alternative is to use the heat 'evolved' by the oxidation to be 'absorbed' (note the sign change) in the second stage. Either method gives $\int_{298}^{T} \Sigma C_P \, dT = -\Delta H$, which must then be solved for the upper temperature.

The reaction is $CH_4 + 2O_2$ $(+8N_2) = CO_2 + 8N_2 + 2H_2O$. Therefore for the combustion products

$$C_P = C_P(CO_2) + 8C_P(N_2) + 2C_P(H_2O)$$

so that (with $T' = T/\text{K}$)

$$\int_{298}^{T} [328.6 + 44.2 \times 10^{-3}T' + 10.5 \times 10^{-6}(T')^2]\,\mathrm{d}T'$$

$$\therefore [328.6T' + 22.1 \times 10^{-3}(T')^2 + 3.5 \times 10^{-6}(T')^3]_{298}^{T} = 890\,000$$

or evaluating the lower limit

$$328.6T' + 22.1 \times 10^{-3}(T')^2 + 3.5 \times 10^{-6}(T')^3 = 890\,000 + 100\,000$$

This equation can be solved by trial and error or graphically. $T' = 2000$ gives 773 000, $T' = 2500$ gives 1014 000, which by linear interpolation gives 2450, yielding 989 200, which is close enough. Thus the adiabatic flame temperature of methane in air is **2450 K**.

A1.10 An air mass moving up to lower air pressure will cool down as a consequence of the work of expansion it is doing at the expense of its enthalpy (if the process is adiabatic).

For an adiabatic expansion of an ideal gas, $\mathrm{d}H = V\mathrm{d}P$ (equation 1.7), and since $\mathrm{d}H = C_P\,\mathrm{d}T$, substitution in the given equation gives

$$\mathrm{d}P = g\rho\,\mathrm{d}z = C_P/V\,\mathrm{d}T$$

or

$$\mathrm{d}T/\mathrm{d}z = V\rho g/C_P$$

but $V\rho = M$ and

$$C_P/M = c_P = 1\ \text{kJ K}^{-1}\,\text{kg}^{-1}$$

$$\therefore\ \mathrm{d}T/\mathrm{d}z = g/c_P = 9.81/1000\ \text{K m}^{-1},$$

or roughly

1 K per 100 m of height

[For moist air, the effective heat capacity would increase considerably since water droplets would form on cooling, liberating $\Delta_{\text{vap}}H$; the 'moist' adiabatic lapse rate would therefore be much smaller.]

Chapter 2

The Second Law of Thermodynamics

2.1 SECOND LAW STATE FUNCTIONS

Historically, the Second Law arose out of work by S. Carnot on the maximum efficiency of heat engines. This led to a statement that all reversible heat engines working in cycles between two temperatures, ΔT apart, must have the same efficiency, $w/q = \Delta T/T$, where q is the heat absorbed at the higher temperature and T is what is now known as the *absolute* or *thermodynamic temperature*. This has many applications in heat engines, heat pumps, and refrigerators.

For chemical thermodynamics, the numerous statements of the Second Law are best formalized in terms of the three state functions S, A, and G:

S, the *entropy*, is defined by

$$\mathrm{d}S = \delta q_{\mathrm{rev}}/T \qquad (2.1)$$

where q_{rev} is reversibly absorbed heat, and the d in dS indicates an exact differential (Appendix A, equation A.5).

A, the *Helmholtz energy* (or *function*), is defined as

$$A = U - TS$$

G, the *Gibbs energy*, or *free energy*, is defined as

$$G = H - TS \qquad (2.2)$$

These definitions lead to three statements of the Second Law, in which the equals sign denotes *reversibility* or *equilibrium* and the $<$ or $>$ sign indicates the requirement for a *spontaneous* (*e.g.* the melting of ice above the freezing point) or *thermodynamically feasible* reaction or change (*e.g.* the explosion of a mixture of hydrogen and oxygen). (Note, however, that all reactions need also to have a path or mechanism, without which they cannot take place at all.) These three statements are:

$$\mathrm{d}S \geqslant 0 \text{ in an isolated system}$$

$$\mathrm{d}A \leqslant 0 \text{ for a system at constant } T \text{ and } V$$

$$\mathrm{d}G \leqslant 0 \text{ for a system at constant } T \text{ and } P \tag{2.3}$$

2.2 REACTIONS AT CONSTANT TEMPERATURE AND PRESSURE

In this context, constant T and P should not be read as constant throughout the reaction, but merely that the system returns to the same T and P it had initially.

In the majority of practical cases we work in a constant pressure environment (*i.e.* at atmospheric pressure), for which the third of these inequalities applies. For a finite change to be feasible, therefore, ΔG must be negative or, in the limiting case of equilibrium, zero. ΔG is a measure of the maximum amount of energy available for 'useful' work (work other than work of expansion). This maximum is only achieved if the work is carried out reversibly, as in some electrochemical cells.

For a reaction at constant T, $\Delta G = G_\mathrm{f} - G_\mathrm{i}$, where f and i stand for final and initial states (or products and reactants) with G_f and G_i given by equation 2.2, so that

$$G_\mathrm{f} = H_\mathrm{f} - TS_\mathrm{f}$$

$$G_\mathrm{i} = H_\mathrm{i} - TS_\mathrm{i}$$

$$\therefore \Delta G = G_\mathrm{f} - G_\mathrm{i} = H_\mathrm{f} - H_\mathrm{i} - T(S_\mathrm{f} - S_\mathrm{i}) = \Delta H - T\Delta S \tag{2.4}$$

Simple differentiation of the defining equation 2.2 will give

$$\mathrm{d}G = \mathrm{d}H - T\mathrm{d}S - S\mathrm{d}T$$

This can be combined with equation 2.1, and substituting equation 1.8 for δq gives the key equation in chemical thermodynamics:

$$dG = V dP - S dT \tag{2.5}$$

Since equation 2.5 may be written down for both the final and initial states:

$$dG_f = V_f dP - S_f dT$$

and

$$dG_i = V_i dP - S_i dT$$

By subtraction

$$d\Delta G = \Delta V dP - \Delta S dT \tag{2.6}$$

It follows from equation A.6 (Appendix A) for state functions that

$$(\partial G/\partial P)_T = V \tag{2.7a}$$

and

$$(\partial \Delta G/\partial P)_T = \Delta V \tag{2.7b}$$

Also

$$(\partial G/\partial T)_P = -S \tag{2.8a}$$

and

$$(\partial \Delta G/\partial T)_P = -\Delta S \tag{2.8b}$$

Substitution in equations 2.2 and 2.4 respectively gives

$$G = H + T(\partial G/\partial T)_P \tag{2.9a}$$

and

$$\Delta G = \Delta H + T(\partial \Delta G/\partial T)_P \tag{2.9b}$$

Rearranging equation 2.9, dividing by $-T^2$, and using $d(u/v) = du/v - u\,dv/v^2$ gives

$$[\partial/\partial T(G/T)]_P = -H/T^2 \tag{2.10a}$$

and

$$[\partial/\partial T(\Delta G/T)]_P = -\Delta H/T^2 \tag{2.10b}$$

Equations 2.9 and 2.10 are referred to as Gibbs–Helmholtz equations.

For a reversible electrochemical cell ΔG can be determined directly from e.m.f. measurements, since all the available chemical energy is converted into electrical work, which is charge × potential difference, where the charge is $n \times \mathcal{F}$ (for 1 mole of particles carrying n unit charges) and $\mathcal{E}$ is the potential difference or e.m.f. Taking the thermodynamic sign convention into account (see equation A.8), this gives

$$\Delta G = -n\mathcal{F}\mathcal{E}$$

and as functions of T

$$\Delta S = -\partial\Delta G/\partial T = n\mathcal{F}\partial\mathcal{E}/\partial T \tag{2.11}$$

Substitution into equation 2.9 gives

$$\Delta H = \Delta G - T\partial\Delta G/\partial T = -n\mathcal{F}(\mathcal{E} - \partial\mathcal{E}/\partial T) \tag{2.12}$$

From equation 2.1 written as a differential w.r.t. T at constant P (see equation A.6), and remembering that $(\partial q/\partial T)_P = (\partial H/\partial T)_P = C_P$ (see equation 1.10):

$$(\partial S/\partial T)_P = C_P/T \tag{2.13a}$$

and

$$(\partial\Delta S/\partial T)_P = \Delta C_P/T \tag{2.13b}$$

2.3 A GEOMETRICAL INTERPRETATION OF ΔG

Most of the above equations find ready application when we consider the 'Ellingham diagram'[3] for the Gibbs energy of formation of metal oxides as a function of temperature (Figure 2.1). (Note that all reactants and products will be in their standard states; in particular, all gases are at the standard pressure.) The following features of this diagram should be noted:

(a) Most of the oxidation reactions shown are feasible since ΔG is negative.
(b) All equations are normalized to 1 mole of oxygen; they can therefore be readily subtracted from one another.
(c) Most individual graphs consist of straight-line sections with changes of slope at phase transitions.
(d) For most lines the slopes at 300 K are roughly the same.

The slope of a ΔG *vs.* T plot is, of course, $(\partial\Delta G/\partial T)_P = -\Delta S$ (see equation 2.7 and Appendix A). A constant slope further implies ΔS independent of T or $\Delta C_p = 0$ (see equation 2.13).

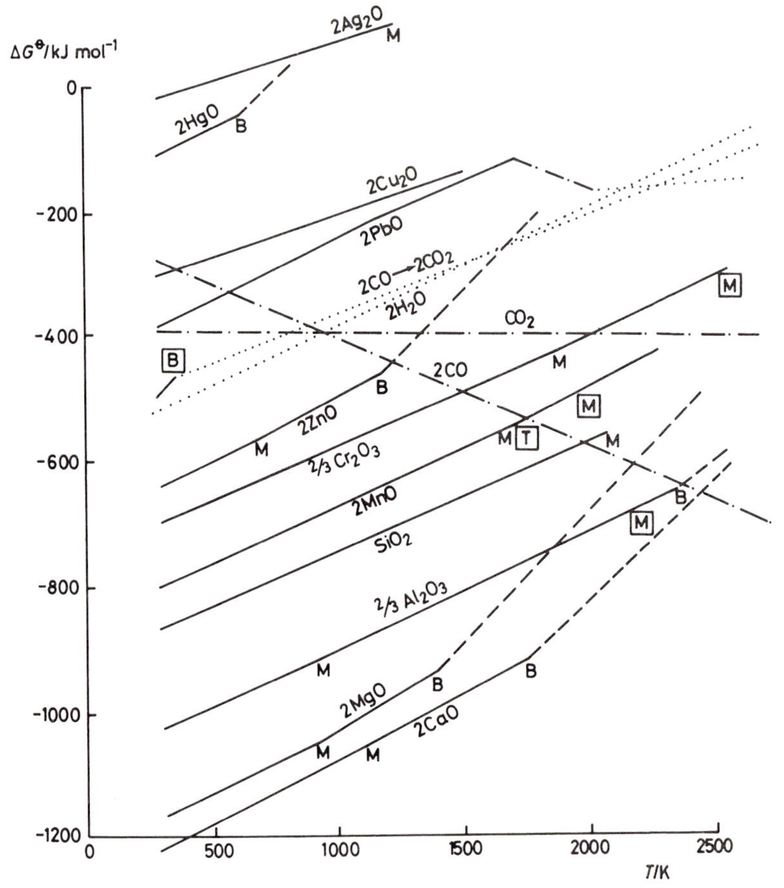

Figure 2.1 *The Ellingham diagram*[3] ($\Delta_f G$ *per mole of* O_2). ΔG *for oxide formation is plotted vs.* T *for one mole of* O_2 *in each case, so that the level difference between two different oxidation reactions represents* ΔG *for a reaction competing for oxygen* (e.g. $2MnO + 2C = 2CO + 2Mn$) *and the cross-over* ($\Delta G = 0$) *marks its equilibrium temperature. Individual plots closely follow straight-line sections with changes at transition points T, at melting points M, and at boiling points B. Unboxed letters refer to the elements, boxed letters to the oxides*

(Actually, ΔC_P is generally a small non-zero quantity and the straight-line relationship should only be remembered as a very useful approximation for any temperature range in which no reactant or product undergoes a phase transition.) Each straight-line section is therefore of the form $\Delta G = \Delta H - T\Delta S$ with an intercept ΔH at $T = 0$ and a slope of $-\Delta S$. (This would still be true of the tangent to ΔG at any temperature if ΔG does not follow a straight line.) Each transition will be accompanied by a *heat of transition*, $\Delta_{tr}H$, which can be read off as the difference in the intercepts, and since phase transitions like melting and boiling are almost perfectly reversible, integration of equation 2.1 will give $\Delta_{tr}S = \Delta_{tr}H/T$, which also determines the change in slope at the transition point.

Note that $S_{gas} \gg S_{liq.} > S_{solid}$ (since S is correlated with 'disorder'), so that any change in the direction solid to liquid to vapour must have a positive entropy (and therefore heat) of transition. For any reaction $\Delta S = S_{products} - S_{reactants}$, and again any gas will make the outstanding contribution to the overall entropy of a reaction so that the disappearance of one mole of gas in the reaction will explain the observation (d) above. The most important exceptions to it are the reactions $C + O_2 = CO_2$, for which there is no change in the amount of gas and $\Delta S \approx 0$, and $2C + O_2 = 2CO$, for which there is an increase of one mole of gas during the reaction, giving a ΔS (and therefore a slope) equal in magnitude but opposite in sign to the majority of reactions.

The normalization to 1 mole of oxygen makes it easy to find ΔG or ΔH for reactions in which one element competes for the oxygen in the oxide of another element. Straight subtraction of the equations makes the elementary oxygen drop out, *e.g.* for the reaction of aluminium with silica we obtain $SiO_2 + 4/3Al = 2/3Al_2O_3 + Si$, and the corresponding subtraction of the ΔG values for any temperature will give ΔG for the reaction – geometrically, subtraction just means measuring the separation of the ΔG levels – and likewise for the ΔH intercepts. In this example the ΔG level drops at all temperatures, giving a negative ΔG and therefore a feasible result over the whole range of temperatures. It may also be seen that carbon will reduce most oxides at sufficiently high temperatures.

It should be noted that ΔH and ΔS are only approximately constant and that ΔG is therefore only approximately linear with T, but any corrections for ΔC_P are unlikely to make a lot of difference to the feasibility of a reaction unless ΔG is within 5 kJ of zero. If more accurate values are required, error limits on the observations, on which the C_P equations are based, need also to be considered.

2.4 MAXWELL RELATIONS

The fact that G is a state function, or equation 2.5 is a total differential, has other implications. As explained in equation A.6 (Appendix A), cross-differentiation leads to the equation

$$(\partial V/\partial T)_P = -(\partial S/\partial P)_T \tag{2.14}$$

one of many 'Maxwell relations' which often lead to useful substitutions. (The left-hand side of the equation is readily determined, the right-hand side is not.) It can be used in the evaluation of the pressure dependence of the enthalpy. From the combined First and Second Law equations (see equation 1.8)

$$\delta q = T\mathrm{d}S = \mathrm{d}H - V\mathrm{d}P \tag{2.15}$$

Therefore, by equation (2.14)

$$T(\partial S/\partial P)_T = (\partial H/\partial P)_T - V = -T(\partial V/\partial T)_P$$

which can be rearranged to the 'generalized equation of state'

$$(\partial H/\partial P)_T = V - T(\partial V/\partial T)_P \tag{2.16}$$

If equation 2.15 is rewritten as $\mathrm{d}H = T\mathrm{d}S + V\mathrm{d}P$, it gives the Maxwell relation

$$(\partial T/\partial P)_S = (\partial V/\partial S)_P$$

2.5 THE THIRD LAW

Brief mention should be made of the third law of thermodynamics, which postulates zero entropy for pure, perfectly crystalline solids at 0 K. Any kind of imperfections in the crystal – such as those in ice, CO, and N_2O – will result in a residual entropy at 0 K, as will any mixture, even the mixture of isotopes of any element; but, provided no unmixing occurs with increase in temperature, or during a reaction, the constant $\Delta_{\mathrm{mix}}S$ can be ignored.

The Third Law is a means of calculating entropies from thermal data (C_P, $\Delta_{\mathrm{tr}}H$) giving

$$S_T = \int_0^T C_P\,\mathrm{d}(\ln T) + \Sigma\,\Delta_{\mathrm{tr}}H/T_{\mathrm{tr}} \tag{2.17}$$

and allows entropies to be obtained from spectroscopic data, using the methods of statistical thermodynamics.

EXAMPLES

2.1 What would be the maximum heat output for a 1 kW heat pump working between the temperatures of 12 and 27 °C?

2.2 Calculate the entropy change in the isothermal reversible expansion of 1 mole of an ideal gas from 2.2 to 22 l.

2.3 Find an expression for $\Delta G^{\ominus}$ as a function of T for the reaction $CCl_4 + O_2 = CO_2 + 2Cl_2$ from the data for S and $\Delta_f H$ below for 298 K but assumed independent of T. Are your results compatible with the use of CCl_4 as a fire extinguisher medium?

	CCl_4	O_2	CO_2	Cl_2
$S^{\ominus}$/J K^{-1} mol^{-1}	215	205	213	223
$\Delta_f H$/kJ mol^{-1}	−137	0	−394	0

2.4 From the Ellingham diagram (Figure 2.1) estimate the following:
(a) The free energy change at 1000 °C for the reaction $4Al + 3O_2 = 2Al_2O_3$.
(b) The entropy changes at 1000 and 1500 °C for the reaction $2Mg + O_2 = 2MgO$.
(c) The free energy, heat, and entropy changes for each of the reactions $Si + O_2 = SiO_2$, $2C + O_2 = 2CO$, and $SiO_2 + 2C = Si + 2CO$ at 1000 °C.
(d) The equilibrium temperature for the reaction $MnO + C = Mn + CO$.

2.5 From the data for 1200 K given below, calculate the information required to decide whether CO or CO_2 is the preferred end-product of the reaction of carbon with steam, and calculate its heat balance (ΔH of the reaction). Suggest a method for maintaining the temperature of the reacting carbon bed by some other reaction with a compensating heat balance.

	H_2	O_2	C	CO	CO_2	H_2O
$S^{\ominus}$/J K^{-1} mol^{-1}	171.7	249.9	28.5	241.0	279.3	240.3
$\Delta_f H$/kJ mol^{-1}	0	0	0	−113.2	−395.0	−248.9

2.6 Explore the thermodynamics of the conversion of graphite into diamond for any temperature, and for 298 K at high pressures.

For C(diamond) = C(graphite), $\Delta G^{\ominus}_{298} = -2.866$ kJ mol^{-1}

For diamond, $S^{\ominus} = 2.44$ J K^{-1} mol^{-1} and $\rho = 3.51$ g cm^{-3}

For graphite, $S^{\ominus} = 5.69$ J K^{-1} mol^{-1} and $\rho = 2.22$ g cm^{-3}

(Entropy and density may be assumed to be independent of T and P.)

2.7 What is the overall 'cell reaction' of the cell

$$\mathrm{Cd}|\mathrm{CdSO_4 soln.}|\mathrm{Cd\ amalgam\ (10\%\ Cd)?}$$

The e.m.f. $\mathcal{E}/\mathrm{mV} = 51.6 + 0.242(t/\mathrm{K} - 20)$, with $t = T - 273.15$ K.

Calculate ΔG, ΔS, and ΔH for the cell reaction at 20 °C.

2.8 Appropriate electrochemical cells gave the following measurements:

$\mathrm{Pb} + \mathrm{I_2} = \mathrm{PbI_2}$,

$\mathcal{E}_{298} = 899.32$ mV,

$\mathrm{d}\mathcal{E}/\mathrm{d}T = -0.042 \pm 0.005$ mV K^{-1}

$\mathrm{Pb} + 2\mathrm{AgI} = \mathrm{PbI_2} + 2\mathrm{Ag}$,

$\mathcal{E}_{298} = 213.5$ mV,

$\mathrm{d}\mathcal{E}/\mathrm{d}T = -0.188 \pm 0.002$ mV K^{-1}

Calculate ΔG, ΔS, and ΔH for the reaction $\mathrm{Ag} + 1/2\mathrm{I_2} = \mathrm{AgI}$ and compare them with the calorimetrically obtained values $\Delta H_{298} = -61.97$ kJ mol^{-1} and $\Delta S_{298} = 13.0 \pm 2.5$ J K^{-1} mol^{-1}.

2.9 From the following data calculate $\Delta G^{\ominus}$ for the reaction $\mathrm{Ni} + 4\mathrm{CO} = \mathrm{Ni(CO)_4(g)}$ at 25 and 225 °C and estimate the effect of raising both the initial and the final pressure to 100 atm at the higher temperature. Comment on the direction the reaction might take at that temperature.

	$\Delta_f H_{298}$/J mol^{-1}	$S^{\ominus}_{298}$/J K^{-1} mol^{-1}	C_P/J K^{-1} mol^{-1}
Ni	0	30.1	$17.0 + 29.5 \times 10^{-3}\,T/\mathrm{K}$
CO	−110 400	198.0	$28.3 + 4.14 \times 10^{-3}\,T/\mathrm{K}$
$\mathrm{Ni(CO)_4}$	−633 900	405.8	$112.1 + 112.0 \times 10^{-3}\,T/\mathrm{K}$

2.10 Show that for an ideal gas $(\partial H/\partial P)_T = 0$. What value has it for a gas for which $V = RT/P + b - a/RT$ (with a and b independent of T)?

ANSWERS

A2.1 The efficiency, w/q, of a reversible heat engine $= \Delta T/T$. In this case, both the work done and the heat absorbed are negative, and $\Delta T/T = 15/300$, so that the heat given out at the higher temperature is $q \times T/\Delta T = 20 \times 1$ kW = **20 kW**.

This does not conflict in any way with the First Law: 1 kW s (= 1 kJ) would be converted without loss from electrical into heat energy, 19 kW s (= 19 kJ) would be absorbed at the lower temperature and given off at the higher temperature.

In actual applications, such high efficiencies are unobtainable, but considerable savings can be achieved with heat pumps.

A2.2 Integration of equation 2.1 at constant T (isothermal!) gives $\Delta S = q_{rev}/T$. It was shown in equation 1.17 that the heat absorbed, q, in an isothermal reversible expansion $= RT \ln(V_f/V_i)$.

$$\therefore \Delta S = R \ln(22/2.2) = \mathbf{19.1\ J\ K^{-1}\ mol^{-1}}$$

A2.3 In the absence of phase changes $\Delta G_T = \Delta H - T\Delta S$, where $\Delta H^{\ominus} = \Sigma\, \Delta_f H$ and $\Delta S^{\ominus} = \Sigma\, S^{\ominus}$, both summed over products minus reactants.

$$\therefore \Delta H^{\ominus}/\text{kJ} = -394 - (-137) = -257$$

and

$$\Delta S^{\ominus}/\text{J K}^{-1} = 2 \times 223 + 213 - 205 - 215 = 239$$

$$\therefore \Delta G^{\ominus}/\text{kJ} = -257 - 0.239T/\text{K}$$

(*N.B.* $\Delta S^{\ominus}$ must be converted from J into kJ)

$\Delta G^{\ominus}$ is negative at all T, so the reaction is thermodynamically feasible, but it is kinetically inhibited by the C atom in CCl_4 being shielded by the surrounding Cl atoms (in the absence of a reactive metal). CCl_4 therefore does not burn and was extensively used in fire extinguishers before its health hazards were recognized.

A2.4 *N.B.* The accuracy of the graph would not justify results better than to the nearest 10 kJ (so ignore solid–solid transitions).

(a) **−2500 kJ** (three times the amount of substance used on the diagram!)
(b) $-\Delta S = (\partial \Delta G/\partial T)_P$, the slope of the straight-line portion. Therefore at 1000 °C, $\Delta S^\ominus = \mathbf{-230\,J\,K^{-1}}$, and $\Delta S_{1500} = \mathbf{-420\,J\,K^{-1}}$.
(c) The $\Delta H^\ominus$ value should be obtained by extrapolation to 0 K. For the reaction $SiO_2 + 2C = Si + 2CO$, direct measurements (from the separation of the two corresponding product and reactant lines) or the subtraction of the individual results should give identical results within the limits of accuracy.

for 1 mol:	$\Delta H^\ominus$/kJ	$\Delta G^\ominus$/kJ	$\Delta S^\ominus$/J K^{-1}
$Si + O_2 = SiO_2$	**−850**	**−670**	**−180**
$2C + O_2 = 2CO$	**−220**	**−400**	**+180**
$SiO_2 + 2C = Si + 2CO$	**+630**	**+270**	**+360**

(d) $\Delta G^\ominus = \Sigma \Delta_f G$ (products minus reactants) = 0 for equilibrium. The equilibrium T must therefore be the T at which products and reactants have equal $\Delta_f G$; this is at **1690 K** (approximately).

A2.5 Thermodynamically, the preferred reaction is the one with the lower ΔG, found as in A2.1 above from ΔH and ΔS. For 1 mol:
(a) $C + H_2O = CO + H_2$, $\Delta H^\ominus = 135.7$ kJ, $\Delta S^\ominus = 143.9\,J\,K^{-1}$, $\Delta G^\ominus = -36.98$ kJ
(b) $C + 2H_2O = CO_2 + 2H_2$, $\Delta H^\ominus = 102.8$ kJ, $\Delta S^\ominus = 113.6\,J\,K^{-1}$, $\Delta G^\ominus = -33.52$ kJ
and (a) is therefore the preferred reaction. However, it is endothermic and would lead to the bed of coke getting cold. To maintain the temperature air is blown in, giving the exothermic reaction (again preferred) $2C + O_2 = 2CO$ with $\Delta H = -226.4$ kJ at 1200 K. The resultant mixture is known as 'water-gas' or 'synthesis gas'.

A2.6 By integration of equation 2.8

$$\Delta G_T = \Delta G_{298} - \int_{298}^{T} \Delta S^{\ominus}\, \mathrm{d}T$$

For graphite to diamond, therefore (with $T' = T/\mathrm{K}$):

$$\Delta G_T^{\ominus}/\mathrm{kJ\,mol^{-1}} = 2866 + 3.25(T' - 298)$$

$$= 1900 + 3.25T'$$

(all positive)

The variation of ΔG with P is given by

$$\Delta G_P = \Delta G^{\ominus} + \Delta V(P - P^{\ominus})$$

(see equation 2.7)

For 1 mole:

$$\Delta V/\mathrm{cm^3} = 12(3.51^{-1} - 2.22^{-1}) = -1.98$$

For equilibrium at 298 K, $\Delta G_{298} = 0$. Therefore

$$P - P^{\ominus} = (\Delta G_P - \Delta G^{\ominus})/\Delta V$$

$$= (0 - 2866)/(-1.98 \times 10^{-6})\ \mathrm{Pa}$$

$$\approx \mathbf{1.45\ MPa}$$

(A higher T and a catalyst are needed to facilitate the reaction.)

A2.7 Both electrodes are reversible to Cd^{2+} ions in the $CdSO_4$ solution, allowing Cd to transfer from one electrode to the other. The cell reaction and therefore the driving force for this cell is the solution of 1 mol of Cd in the 10% Cd amalgam.

$$\Delta G = -2 \times 96\,487 \times 51.6 \times 10^{-3} = \mathbf{-9.96\ kJ\,mol^{-1}}$$

(see equations 2.11 and 2.12)

$$\Delta S = 2 \times 96\,487 \times 0.242 \times 10^{-3} = \mathbf{46.7\ J\,K^{-1}\,mol^{-1}}$$

($T = t + 273$ K; $\mathrm{d}T = \mathrm{d}t$)

$$\Delta H = \Delta G + T\Delta S = -9.96 + 0.0467 \times 293$$

$$= \mathbf{3.72\ kJ\,mol^{-1}}$$

A2.8 *N.B.* The e.m.f. is an *intensive* property, *i.e.* independent of the amount, whereas ΔG is *extensive*, or proportional to the amount; therefore subtracting the second equation (and its e.m.f.) from the first will give the correct e.m.f., but the value of n in $\Delta G^{\ominus} = -n\mathcal{F}\mathcal{E}$ will be 2 for $2Ag + I_2 = 2AgI$ and 1 for the required reaction. Therefore

$$\Delta G^{\ominus} = -\mathcal{F}\mathcal{E} = -96\,487(0.6858 \pm 2)10^{-3}$$

$$= \mathbf{-66.17 \pm 0.02\ kJ\ mol^{-1}}$$

$$\Delta S^{\ominus} = \mathcal{F}\partial\mathcal{E}/\partial T = 96\,487(0.146 \pm 7)10^{-3}$$

$$= \mathbf{14.1 \pm 0.7\ J\ K^{-1}\ mol^{-1}}$$

$$(\Delta S^{\ominus}_{cal} = 13.0 \pm 2.5\ J\ K^{-1}\ mol^{-1})$$

(*N.B.* Error limits must be added for additions and subtractions.)

$$\Delta H^{\ominus} = \Delta G^{\ominus} + T\Delta S^{\ominus} = \mathbf{-61.97 \pm 0.2\ kJ\ mol^{-1}}$$

$$(\Delta H^{\ominus}_{cal} = -61.97\ kJ)$$

Agreement is well within the error limits ($\Delta G^{\ominus}_{cal} = -65.8 \pm 0.8$ kJ).

A2.9 $\Delta G = \Delta H - T\Delta S$, so that

$$\Delta G^{\ominus}_{298} = -192.3 + 0.4163 \times 298 = \mathbf{-68.2\ kJ\ mol^{-1}}$$

$$\Delta H^{\ominus}_{498} = \Delta H^{\ominus}_{298} + \int_{298}^{498} \Delta C_P\, dT$$

and

$$\Delta S^{\ominus}_{498} = \Delta S^{\ominus}_{298} + \int_{298}^{498} \Delta C_P\, d\ln T$$

with $\Delta C_P = -18.1 + 66 \times 10^{-3}T\ J\ K^{-1}\ mol^{-1}$ [remember $d\ln T = dT/T$]

$$\therefore\ \Delta H^{\ominus}_{498} = -192.3 + 1.6 = -190.7\ kJ\ mol^{-1}$$

and

$$\Delta S^{\ominus}_{498} = -416.3 - 18.1\ln(498/298) + 0.066 \times 200$$

$$= -412.4\ J\ K^{-1}\ mol^{-1}$$

$$\therefore\ \Delta G^{\ominus}_{498} = \mathbf{14.7\ kJ\ mol^{-1}}$$

It is worth noting that the simplified calculation $\Delta G^{\ominus}_{498} = \Delta H^{\ominus}_{298} - T\Delta S^{\ominus}_{298}$ (which should always be used for first approximations) gives 15.0 instead of 14.7 kJ – hardly a significant difference!

The reaction is therefore seen to be feasible at 298 K and the reverse reaction is feasible at 498 K in the standard state of each reactant and product.

Increasing the pressure from 1 atm to 100 atm for each gas (Ni will not be much affected), $\Delta V = \Delta nRT/P$ (if ideal behaviour is assumed). But $\partial \Delta G/\partial P = \Delta V$ (equation 2.7) and Δn for the reaction is -3 moles of gas. Integration therefore gives

$$\Delta G^{100} = \Delta G^{\ominus} - 3RT\ln 100 = \mathbf{-42.5\ kJ\ mol^{-1}}$$

Therefore application of pressure makes the forward reaction feasible.

[Similar calculations underlie the industrial refining of nickel.]

A2.10 $(\partial H/\partial P)_T = V - T(\partial V/\partial T)_P$ (equation 2.16)
Substituting $V = RT/P$ for an ideal gas:

$$(\partial H/\partial P)_T = RT/P - T(R/P) = 0$$

For the equation of state $V = RT/P + b - a/RT$:

$$\partial V/\partial T = R/P + a/RT^2$$

and

$$\therefore \boldsymbol{(\partial H/\partial P)_T = b - 2a/RT}$$

Chapter 3

Ideal Gas Equilibrium

In ideal gas mixtures, the pressure exerted by each gas i (its *partial pressure*, p_i) is independent of other gases present and is calculated from the ideal gas equation as if it alone occupied the space available. For n_i moles of any gas in such a mixture

$$\Sigma p_i = \Sigma n_i RT/V = P \text{ or } P_t \text{ (the total pressure)} \qquad (3.1)$$

and

$$p_i = n_i RT/V = c_i RT = x_i RT/V = x_i P_t \qquad (3.2)$$

since n_i/V is the concentration, c_i, and x_i (or y_i) is the mole fraction of the gas in the mixture.

Following the 'equilibrium box' treatment of van't Hoff,[4] the Gibbs energy of an ideal gas reaction can best be worked out in three stages:

(1) The expansion of each reactant (denoted by the subscript re) to its equilibrium pressure.
(2) The reaction at equilibrium pressures (p_e).
(3) The expansion of each product (subscript pr) from its equilibrium to its final pressure.

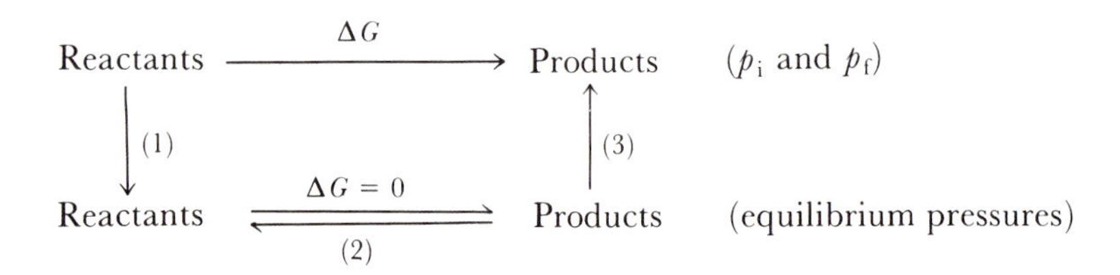

Since $\Delta G = 0$ at equilibrium, stage 2 does not contribute to

ΔG of the overall reaction. Stages 1 and 3 can both be found from equation 2.7, with $V = n_i RT/p_i$ for each gas. For chemical reactions it is usual to replace n_i by the stoichiometric coefficients ν_i and to call the overall change in the amount of gas $\Delta\nu$.

The standard Gibbs energy change for a reaction, $\Delta G^\ominus$, is the ΔG when each product and reactant is at the standard pressure $P^\ominus$, so that each initial and final partial pressure (p_i and p_f, respectively) $= P^\ominus = 10^5$ Pa. In this case ΔG for stages 1 and 3 will add up to $\Delta G^\ominus$.

$$\therefore \Sigma\,[\nu_\mathrm{re} RT \ln (p_\mathrm{e}/P^\ominus)] - \Sigma\,[\nu_\mathrm{pr} RT \ln (p_\mathrm{e}/P^\ominus)]$$
$$= -\Sigma\,[\nu RT \ln (p_\mathrm{e}/P^\ominus)] = \Delta G^\ominus$$

[Σ(products minus reactants) (equivalent to making all ν_re negative)]
This can be contracted further by writing sums of logs as logs of products, so that

$$\Delta G^\ominus = -RT \ln \Pi (p_\mathrm{e}/P^\ominus)^\nu = -RT \ln K^\ominus \tag{3.3}$$

In this expression $K^\ominus$, the *standard equilibrium constant*, is the product of the equilibrium partial pressures with exponents as shown for the example reaction $N_2 + 3H_2 = 2NH_3$, for which

$$K^\ominus = \frac{(p_\mathrm{e}/P^\ominus)^2_{\mathrm{NH_3}}}{(p_\mathrm{e}/P^\ominus)_{\mathrm{N_2}}(p_\mathrm{e}/P^\ominus)^3_{\mathrm{H_2}}} = K_P/(P^\ominus)^{\Delta\nu}$$

Through its relationship to $\Delta G^\ominus$, $K^\ominus$ is clearly a function of T and the standard pressure $P^\ominus$, but not the total pressure P.

Where any of the initial or final pressures differ from the standard pressure, $P^\ominus$ in equation 3.3 would have to be replaced by the value of the corresponding initial and final pressure, p'. The resulting ratio can be recast into the form

$$p_\mathrm{e}/p' = (p_\mathrm{e}/P^\ominus)/(p'/P^\ominus)$$

The general equation for ΔG then becomes

$$\Delta G = -RT \ln \Pi (p_\mathrm{e}/P^\ominus)^\nu + RT \ln \Pi (p'/P^\ominus)^\nu$$

or

$$\Delta G = -RT \ln K^\ominus + RT \ln \Pi (p'/P^\ominus)^\nu \tag{3.4}$$

Equation 3.4 is also known as the van't Hoff isotherm.

It should be noted that the argument of a logarithm must be a pure number. In some textbooks K_P is used for $K^\ominus$. In that case the dimensionless quantity $K_P/(P^\ominus)^{\Delta\nu}$ is intended, but with $P^\ominus$ equal to the standard atmosphere and not the bar (10^5 Pa), which replaced it as the standard pressure in 1982.[5]

Substitution of equation 3.2 into the expression for K_P gives similar expressions in terms of concentrations or mole fractions, so that

$$K^\ominus(P^\ominus)^{\Delta\nu} = K_P = K_c(RT)^{\Delta\nu} = K_x P^{\Delta\nu} \qquad (3.5)$$

Since $K^\ominus$ and K_P are independent of P, so (by equation 3.5) is K_c, while K_x varies with P (see equation 3.8 below). (K_c is rarely used for gas reactions, but K_x is required for the calculation of yields.)

It follows from equations 3.3 and 2.10 that

$$\partial \ln K^\ominus/\partial T = \partial/\partial T(-\Delta G^\ominus/RT) = \Delta H^\ominus/RT^2 \qquad (3.6)$$

From equation 3.5, taking logs and rearranging:

$$\ln K^\ominus = \ln K_x + \Delta\nu \ln(P/P^\ominus) \qquad (3.7)$$

with the last term disappearing on differentiating w.r.t. T at constant P, so that equation 3.6 applies equally to $\ln K_x$. Again, if $\Delta\nu = 0$, $K^\ominus$ and K_x become equal and are then both independent of P. But if $\Delta\nu \neq 0$, differentiation of equation 3.7 w.r.t. P gives

$$\partial \ln K^\ominus/\partial P = 0 = \partial \ln K_x/\partial P + \Delta\nu/P \quad [\Delta\nu \ln P^\ominus = \text{constant}]$$

so that

$$\partial \ln K_x/\partial P = -\Delta\nu/P = -\Delta\nu V/PV = -\Delta V/RT \qquad (3.8)$$

Both equations 3.6 and 3.8 can be regarded as quantitative expressions of *Le Chatelier's principle*, *i.e.* a system subjected to a constraint will react so as to minimize that constraint. Thus if it is an increase in temperature, the system will react by absorbing heat, *i.e.* undergo an endothermic process, if possible. To counteract an increase in pressure, the equilibrium will be displaced towards the side having the smaller volume, or amount of gas (if any). This, in turn will affect the mole fraction at equilibrium and therefore the *equilibrium conversion*, defined as the fraction or percentage of any reactant converted into product.

For ideal gas reactions, the effect of any inert gas present will be the same as the lowering of the total pressure and therefore affect the conversion only if $\Delta\nu \neq 0$.

EXAMPLES

(Ideal gas behaviour should be assumed for all these examples.)

3.1 In the gas equilibrium $2SO_2 + O_2 = 2SO_3$ at 1000 K and at a total pressure of $P^\ominus = 1$ atm the analysis of the equilibrium mixture gave the following mole fractions:

$$0.564 \text{ for } SO_2, \quad 0.103 \text{ for } O_2, \quad 0.333 \text{ for } SO_3$$

Calculate $K^\ominus$, $\Delta G^\ominus$, and K_P.

3.2 (a) To what starting and finishing pressures of reactants and products do the results in 3.1 refer and how would ΔG be affected (using the same initial conditions) if the SO_3 is removed from the reactor at 0.1 atm?
(b) What would be the results for $P^\ominus = 1$ bar?

3.3 Using the data given below, estimate $\Delta G^\ominus_{698}$ for the reaction $H_2O(g) + CO = CO_2 + H_2$, stating your assumptions.
If equimolar amounts of CO and H_2O are allowed to reach equilibrium at 698 K, what will be the mole ratio of H_2 to H_2O?
The data are for 298 K and for $P^\ominus = 0.1$ MPa.

	CO_2	CO	$H_2O(g)$
$\Delta_f H$/kJ mol^{-1}	−393.5	−110.5	−241.8
$\Delta_f G$/kJ mol^{-1}	−394.4	−137.2	−228.6

3.4 If a given amount of a 2 : 1, by volume, mixture of NO and O_2 is allowed to reach equilibrium with its reaction product NO_2 at 344 °C and 1 bar, a contraction in volume of 25% is observed. What values of K_x and $K^\ominus$ would account for this contraction? What would you expect them to be at a pressure of 10 bar?

3.5 For the formation reaction of NH_3 at 298 K, $\Delta H^\ominus = -46.0 \pm 0.1$ kJ mol^{-1} and $\Delta S^\ominus = -100.0 \pm 0.5$ J K^{-1} mol^{-1}. Calculate $K^\ominus$ and K_x at 1 and 100 bar.
Would the equilibrium shift with temperature be favourable to the formation of NH_3?

3.6 For the hydrogenation at 100 °C of benzene to cyclohexane:

$$C_6H_6 + 3H_2 = C_6H_{12}, \ \Delta H^\ominus = -192 \text{ kJ mol}^{-1}$$

and the partial pressures at equilibrium were found to be:

$$p(H_2) = 1 \text{ torr},\ p(C_6H_6) = 69 \text{ torr},\ p(C_6H_{12}) = 690 \text{ torr}$$

Calculate the equilibrium constants K_P, $K^{\ominus}$, and $\Delta G^{\ominus}$ for the hydrogenation reaction at 100 °C and estimate K_P at 90 °C.
[1 torr ≈ 133.3 Pa]

3.7 If $\Delta_f G(NO)/J\ mol^{-1} = 90\,400 - 10.5T/K$, estimate the approximate equilibrium partial pressure of NO produced at $P^{\ominus}$ from a 4 : 1 mixture of N_2 and O_2 at 2000 K.

3.8 $N_2O_4 = 2NO_2$ has $\Delta H^{\ominus} = 56.9\ kJ\ mol^{-1}$ and $\Delta S^{\ominus} = 174\ J\ K^{-1}\ mol^{-1}$. If the above data are valid for 400 K, work out the composition of the mixture at equilibrium for total pressures of 1 and 10 bar; what would it be for 1 bar at 350 K?

3.9 From the data given below estimate $\Delta G^{\ominus}$ at 573 K for the reaction $2H_2 + CO = CH_3OH(g)$.
What would be the value of ΔG_{573} if all reactants and products were at a partial pressure of 25 MPa each?
At 298 K:

$$2H_2 + CO = CH_3OH(g),\ \Delta H^{\ominus}/kJ\ mol^{-1} = -90.2$$

	$CH_3OH(g)$	CO	H_2
$S^{\ominus}/J\ K^{-1}\ mol^{-1}$	239.7	197.6	130.6
$C_P/J\ K^{-1}\ mol^{-1}$	43.8	29.14	28.84

3.10 In an adiabatic flow reactor for the oxidation of SO_2 to SO_3, the reacting mixture of SO_2, O_2, and N_2 in a mole ratio of 1 : 1 : 8 enters the reactor at 800 K.
(a) Assuming the continuous reaction goes to completion, estimate the steady-state temperature reached by the reactor.
(b) What further calculations are required to find the extent of the reaction at equilibrium?

Data for 800 K

	SO_3	SO_2	O_2	N_2
$\Delta_f H^{\ominus}/kJ\ mol^{-1}$	−406.1	−307.7	0	0
$\Delta_f G^{\ominus}/kJ\ mol^{-1}$	−321.9	−298.4	0	0
$C_P/J\ K^{-1}\ mol^{-1}$	72.8	52.4	33.8	31.4

ANSWERS

A3.1 The standard pressure is here defined as 1 atm, so that $\Delta G^\ominus$ and $K^\ominus$ would also refer to that standard. Therefore $P/P^\ominus = 1$ and $K_x = K^\ominus$ (equation 3.5).

$$\therefore K^\ominus = 0.333^2/(0.564^2 \times 0.103) = \mathbf{3.38}$$

and

$$K_P = \mathbf{3.38\ atm^{-1}}$$

$$\Delta G^\ominus = -RT \ln K^\ominus = -8314 \ln 3.38 = \mathbf{-10.1\ kJ\ mol^{-1}}$$

A3.2 (a) The starting and finishing partial pressure of each gas is $P^\ominus$. By equation 3.4

$$\Delta G = \Delta G^\ominus + RT \ln (0.1^2)$$

$$= -10.1 + 8314(-4.61)$$

$$= \mathbf{-48.4\ kJ\ mol^{-1}}$$

(b) If the data still refer to 1 atm, $K^\ominus = K_x(P/P^\ominus)^{\Delta\nu}$

$$\therefore K^\ominus = 3.38(1.01325)^{-1} = 3.34$$

and

$$\Delta G^\ominus = \mathbf{-10.0\ kJ\ mol^{-1}}$$

$p/P^\ominus$ would become 0.1013, so that $\Delta G = \mathbf{-48.1\ kJ\ mol^{-1}}$

A3.3 Assuming $\Delta C_P = 0$, which is equivalent to ΔH and ΔS independent of T, the value of ΔG for 698 K can be found from $\Delta G = \Delta H - T\Delta S$. First

$$\Delta S_{698} = \Delta S_{298} = (\Delta H_{298} - \Delta G_{298})/298\ \text{K}$$

$$\Delta_f H^\ominus/\text{kJ mol}^{-1} = -393.5 - (-110.5) - (-241.8)$$

$$= -41.2$$

$$\Delta_f G^\ominus/\text{kJ mol}^{-1} = -394.4 - (-137.2) - (-228.6)$$

$$= -28.6$$

$$\therefore \Delta S^\ominus/\text{J K}^{-1}\text{ mol}^{-1} = -12.6 \times 10^3/298 = -42.3$$

$$\therefore \Delta G^\ominus_{698}/\text{kJ mol}^{-1} = -41.2 - 698(-42.3 \times 10^{-3})$$

$$= -11.7$$

If α moles of H_2O and CO have reacted to form CO_2 and H_2 when equilibrium has been attained:

$$H_2O(g) + CO \ = \ CO_2 + H_2$$

Moles at equilibrium: $1 - \alpha$ $\qquad \alpha \qquad \alpha \qquad$ (total 2)

then $K_x = (\alpha/2)^2/[(1 - \alpha)/2]^2 = [\alpha(1 - \alpha)]^2$
Therefore the mole ratio of H_2 to H_2O at equilibrium is $\alpha/(1 - \alpha) = (K_x)^{1/2}$. But $K_x = K^\ominus$ when $\Delta\nu = 0$, and therefore

$$K_x = \exp(-\Delta G^\ominus/RT)$$

$$= \exp(11\,700/8.314 \times 698) = 2.02$$

Thus the ratio H_2/H_2O is $2.02^{1/2} =$ **1.42**.

A3.4 This is best tackled by considering the extent of reaction when equilibrium is attained. The units could be volume or moles.

	$2NO$ +	O_2 →	$2NO_2$	
Initially	2	1	0	(total 3)
At equilibrium	$2 - 2\alpha$	$1 - \alpha$	2α	(total $3 - \alpha = 2.25$)
(25% of 3 is 0.75)				$\therefore \alpha = 0.75$
$\therefore$ at equilibrium	0.5	0.25	1.50	

Dividing the amounts by the total:

$$K_x = (1.5/2.25)^2/(0.5^2 \times 0.25/2.25^3) = \mathbf{81}$$

Remembering that $K_x = K^\ominus$ if $P = P^\ominus$ (equation 3.5) and that $K^\ominus$ (unlike K_x) is independent of P in ideal gas reactions, $K^\ominus$ = **81 at either *P***.

At 10 bar

$$K_x = K^\ominus(P^\ominus/P)^{\Delta\nu} = \mathbf{810} \text{ (since } \Delta\nu = -1)$$

A3.5 $\Delta G^\ominus = \Delta H^\ominus - T\Delta S^\ominus$

$$= -46.0 \pm 0.1 - 298(-100 \pm 0.5) \times 10^{-3}$$

$$= -16.2 \pm 0.25 \text{ kJ mol}^{-1}.$$

(*N.B.* The error limits are widened by adding their absolute values.)

Working out $K^\ominus$ from $\exp(-\Delta G^\ominus/RT)$ for the values -16.45 and -15.95 gives $K^\ominus = 765$ and 625, respectively.

$$\therefore\ K^{\ominus} = K_x = \mathbf{695 \pm 30} \text{ at 1 bar}$$

From equation 3.5, $K_x = K^{\ominus}(P^{\ominus}/P)^{\Delta\nu}$ with $K^{\ominus}$ independent of the pressure, and $\Delta\nu$ for the formation reaction (1 mole!) is -1, giving

$$K_x = \mathbf{6.95 \pm 0.3 \times 10^4} \text{ for 100 bar or 10 MPa}$$

$$\partial\Delta G/\partial T = -\Delta S = -(-100)\ \text{J K}^{-1}\ \text{mol}^{-1} > 0$$

An increase in T would make ΔG more positive and is therefore **unfavourable** to the reaction.

A3.6 $K_P = 690/(1 \times 69) = \mathbf{10\ torr^{-3}}$.
While K_P can have any pressure units, it must be in terms of $P^{\ominus}$ to be numerically equal to $K^{\ominus}$. By equation 3.5

$$K^{\ominus} = K_P(P^{\ominus})^{-\Delta\nu}$$

but $P^{\ominus} \approx 750$ torr and $\Delta\nu = -3$. $\therefore\ K^{\ominus} = \mathbf{4.22 \times 10^9}$

and

$$\Delta G^{\ominus}_{373} = -RT\ln K^{\ominus} = \mathbf{-68.73\ kJ\ mol^{-3}}$$

Using the integrated form of equation 3.6:

$$\ln(K_2^{\ominus}/K_1^{\ominus}) = -(\Delta H/R)(T_2^{-1} - T_1^{-1})$$

$$= -\Delta H(T_1 - T_2)/RT_1T_2$$

$$\therefore\ \ln K_2^{\ominus} = \ln K_1^{\ominus} + (192 \times 10^4)/(8.314 \times 373 \times 363)$$

$$= 23.87$$

$$\therefore \text{ at } 90\ ^{\circ}\text{C},\ K^{\ominus} = \exp(23.87) = 23.2 \times 10^9$$

$$\therefore\ K_P = \mathbf{23.2 \times 10^9\ bar^{-3}}$$

A3.7 From the given data, $\Delta_f G = 69.4$ kJ mol^{-1} at 2000 K.

$$\therefore\ \ln K^{\ominus} = -69\,400/(8.314 \times 2000) = -4.174$$

$$\therefore\ K^{\ominus} = 0.0154 = K_x \quad [\text{since } \Delta\nu = 0 \text{ (equation 3.7)}]$$

In a 4 : 1 mixture $x(N_2) = 0.8$ and $x(O_2) = 0.2$, giving

$$\tfrac{1}{2}N_2 \quad + \quad \tfrac{1}{2}O_2 \quad = NO$$

$$0.8 - \alpha \quad 0.2 - \alpha \quad 2\alpha \text{ [total 1 (mol or mol fraction)]}$$

$$\therefore\ K_x = x(NO)/[(0.8 - \alpha)(0.2 - \alpha)]^{1/2},$$

$$\text{with } x(NO) = 2\alpha$$

but $\alpha \ll 0.2$ since K_x is small

$$\therefore K_x \approx 2\alpha/(0.8 \times 0.2)^{1/2} = 0.0154$$

so that $2\alpha \approx 0.04K_x = 6.16 \times 10^{-3}$ [neglecting α in $(0.2 - \alpha)$]

If desired, a second approximation can be carried out with $\alpha = 0.003$, giving

$$x(\text{NO}) = (0.797 \times 0.197)^{1/2}K_x = \mathbf{6.10 \times 10^{-3}} = p/P^{\ominus}$$

A3.8 At equilibrium, let $(1 - \alpha)N_2O_4$ remain after formation of $2\alpha NO_2$; the total will then be $(1 + \alpha)$ mol. The mole fractions are then

$$x(\text{N}_2\text{O}_4) = (1 - \alpha)/(1 + \alpha)$$

and

$$x(\text{NO}_2) = 2\alpha/(1 + \alpha)$$

$$\therefore K_x = [2\alpha/(1 + \alpha)]^2/[(1 - \alpha)/(1 + \alpha)]$$

$$= 4\alpha^2/(1 - \alpha)(1 + \alpha) = 4\alpha^2/(1 - \alpha^2)$$

Solving for α^2 gives $K_x/(K_x + 4)$. But from equation 3.5

$$K_x = K^{\ominus}(P^{\ominus}/P)^{\Delta\nu}$$

with $\Delta\nu = 1$ in this case.

$$K^{\ominus} = \exp(-\Delta G^{\ominus}/RT)$$

and

$$\Delta G^{\ominus}/\text{kJ mol}^{-1} = 56.9 - 400 \times 174 \times 10^{-3} = -12.7$$

$$\therefore K^{\ominus} = \exp(12\,700/8.314 \times 400) = 45.55 = K_x$$

for

$$P = P^{\ominus} \text{ (1 bar)}$$

but

$$K^{\ominus} = 10K_x \text{ for } P = 10 \text{ bar}$$

At 1 bar $\alpha = (45.55/49.55)^{1/2} = 0.959$

$$\therefore x(\text{NO}_2) = 2 \times 0.959/(1 + 0.959) = \mathbf{0.979}$$

$$\text{leaving } x(\text{N}_2\text{O}_4) = \mathbf{0.021}$$

At 10 bar $\alpha = (4.56/8.56)^{1/2} = 0.730$

$$\therefore\ x(NO_2) = 1.46/1.73 = \mathbf{0.844}$$

and

$$x(N_2O_4) = \mathbf{0.156}$$

(As predicted by Le Chatelier's principle, less NO_2 is formed.)

For 350 K (see equation 3.6)

$$\ln K^{\ominus}_{350} = \ln K^{\ominus}_{400} - \Delta H^{\ominus}(1/350\ \mathrm{K} - 1/400\ \mathrm{K})/R$$

$$= \ln 45.55 - 56\,900(3.57 \times 10^{-4})/8.314$$

$$= 1.375$$

$$\therefore\ K_{x,350} = K^{\ominus}_{350} = 3.953$$

so that

$$\alpha = (3.953/7.953)^{1/2} = 0.705$$

$$\therefore\ x(NO_2) = 1.410/1.705 = \mathbf{0.827}$$

and

$$x(N_2O_4) = \mathbf{0.173}$$

(Again, an endothermic process is favoured by an increase in T.)

A3.9 From the data

$$\Delta S^{\ominus}_{298}/\mathrm{J\ K^{-1}\ mol^{-1}} = 239.7 - 197.6 - 2 \times 130.6$$

$$= -219.1$$

$$\Delta C_P/\mathrm{J\ K^{-1}\ mol^{-1}} = 43.8 - 29.14 - 2 \times 28.84 = -43.02$$

At 573 K

$$\Delta H^{\ominus} = \Delta H^{\ominus}_{298} + \Delta C_P(573 - 298)\ \mathrm{K}$$

$$= -90.2 + (-43.02 \times 10^{-3}) \times 275$$

$$= -102.03\ \mathrm{kJ\ mol^{-1}}$$

$$\Delta S^{\ominus} = \Delta S^{\ominus}_{298} + \Delta C_P \ln(573/298)$$

$$= -219.1 + (-43.02) \times \ln 1.923$$

$$= -247.2\ \mathrm{J\ K^{-1}\ mol^{-1}}$$

$$\therefore \Delta G^{\ominus}_{573} = -102.0 - 573 \times (-247.2 \times 10^{-3})$$
$$= 39.6 \text{ kJ mol}^{-1}$$

From equations 3.3 and 3.4 it follows that

$$\Delta G = \Delta G^{\ominus} + RT \ln \Pi (p/P^{\ominus})^{v}$$

where the product, Π, is taken over the initial and final pressures of each constituent, here $250P^{\ominus}$ each.

$$\therefore \Delta G/\text{kJ mol}^{-1} = 39.6 + 8.314 \times 10^{-3}$$
$$\times 573 \times \ln(250^{-2}) = \mathbf{-13.0}$$

(The increased pressures make the reaction at 573 K feasible.)

A3.10 (a) For the reaction $SO_2 + \frac{1}{2}O_2 = SO_3$, $\Delta H^{\ominus}$ can be found from the data

$$\Delta H^{\ominus}/\text{kJ mol}^{-1} = -406.1 - (-307.7) = -98.4$$

In an adiabatic flow reactor, all this heat must go into increasing ΣH of the reaction products. At completion these would be $SO_3 + \frac{1}{2}O_2 + 8N_2$ with a $\Sigma C_P/\text{J K}^{-1} = 72.8 + 16.9 + 251.2 = 340.9$.

The temperature of the product gases would therefore increase by $98\,400/340.9 = 289$ K to 1089 K. Initially, some heat would go into heating up the reactor itself, and this would in turn heat up some of the incoming gases, but when final thermal equilibrium is established the heat generated in the reactor, ΔH, must be removed by the product gas. No heat is lost by the reactor itself, since it is adiabatic.

(b) While equilibrium at 800 K will be close to complete conversion ($\Delta G^{\ominus} = -23.5 \text{ kJ mol}^{-1}$), at higher temperatures $\Delta G^{\ominus}$ would increase, and the equilibrium would shift towards the reactants. $\Delta G^{\ominus}$ at the new temperature will be needed.

It will then be necessary to calculate K_x, taking into account the reduction in the effective pressure (see text) due to the presence of inert gases. The mole fractions at equilibrium can then be found as, for instance, in example 3.3.

If only α moles of SO_2 are converted into SO_3, then the amount of heat generated would only be $\alpha(-\Delta H^{\ominus})$. The composition of the equilibrium product and its $\Sigma\, C_P$ would become functions of α.

Several iterations would be required to find the steady-state temperature.

Chapter 4

Phase Equilibrium

4.1 THE CLAUSIUS–CLAPEYRON EQUATION

Phase changes (*e.g.* fusion, evaporation) taking place at constant temperature and pressure have $\Delta G = 0$ (for equilibrium). By equation 2.6, therefore, if we change the temperature and vapour pressure to an adjacent set of equilibrium conditions (so again giving $\Delta G = 0$), we have

$$\mathrm{d}\Delta G = 0 = \Delta V\,\mathrm{d}p - \Delta S\,\mathrm{d}T \quad (4.1)$$

with changes in equilibrium vapour pressure and temperature given by

$$(\mathrm{d}p/\mathrm{d}T)_{\text{phase change}} = \Delta S/\Delta V \quad (4.2)$$

For ΔS we can substitute q_{rev}/T (equation 2.1). The reversibly absorbed heat of a phase change used to be called 'latent heat', but is now simply the enthalpy of transition, $\Delta_{\text{tr}}H$. For specific changes the recommended subscripts are: 'fus' for fusion, 'vap' for vaporization, and 'sub' for sublimation. In general then, equation 4.2 becomes the Clapeyron equation:

$$(\mathrm{d}p/\mathrm{d}T)_{\text{ph ch}} = \Delta_{\text{tr}}H/T\Delta V \quad (4.3)$$

For transitions from a condensed (denoted by subscript cd) phase (solid or liquid) to a gas or vapour phase, further simplifications can be introduced by treating the vapour as an ideal gas and putting $\Delta V = V_{\text{g}} - V_{\text{cd}} \approx V_{\text{g}} \approx RT/p$, since V_{cd} is negligible compared with V_{g}.

With these approximations and rearrangement, equation 4.3 becomes the Clausius–Clapeyron equation:

$$[\mathrm{d}(\ln p)/\mathrm{d}T]_{\text{vap}} = \Delta_{\text{vap}}H/RT^2 \quad (4.4)$$

By substituting $T^{-2}\,dT = -d(T^{-1})$, equation 4.4 can be recast as

$$[d(\ln p)/d(T^{-1})]_{vap} = -\Delta_{vap}H/R \qquad (4.5)$$

indicating that the $\ln p'$ *vs.* $1/T$ plot† will be linear with a slope of $-\Delta_{vap}H/R$ as long as the temperature variation of $\Delta_{vap}H$ can be neglected. From the slope a mean $\Delta_{vap}H$ may be determined.

The integrated form of the equation is used to calculate the vapour pressure and should then be written as a definite integral, usually with the boiling point‡ as the lower limit.

$$\ln\frac{p}{p_0} = -\frac{\Delta_{vap}H}{R}\left(\frac{1}{T} - \frac{1}{T_0}\right) \qquad (4.6)$$

4.2 IDEAL SOLUTIONS

The Clausius–Clapeyron equation (equation 4.4 or equation 4.6) finds extensive application in the study of solutions. Ideal solutions may be defined as those obeying Raoult's law, which may be stated as

$$p_A = x_A p_A^* \qquad (4.7)$$

where p_A^* stands for the vapour pressure of the pure solvent and x_A is its mole fraction in the solution. Where a distinction can be made between solvent and solute it is usual to reserve the subscript 1 or A for the solvent.

A non-volatile solute in a volatile solvent will reduce the vapour pressure in accordance with Raoult's law (equation 4.7) if behaviour is ideal. The standard boiling point will be reached when the vapour pressure (v.p.) reaches the value $P^\ominus$, as shown in Figure 4.1. If equation 4.7 is applied to the solution between the standard boiling points of the pure solvent and the solvent in solution:

$$\ln\frac{xP^\ominus}{P^\ominus} = \ln x = -\frac{\Delta_{vap}H}{R}\left(\frac{1}{T_0} - \frac{1}{T'}\right) = -\frac{\Delta_{vap}H}{R}\frac{(T' - T_0)}{T_0 T'} \qquad (4.8)$$

†p' may be any pressure divided by its unit pressure.

‡Note that the standard pressure is now taken as 1 bar (100 kPa), but used to be the standard atmosphere (101.325 kPa), and most boiling points in the literature relate to 1 atm. Such a b.p. is now referred to as the 'normal b.p.'.

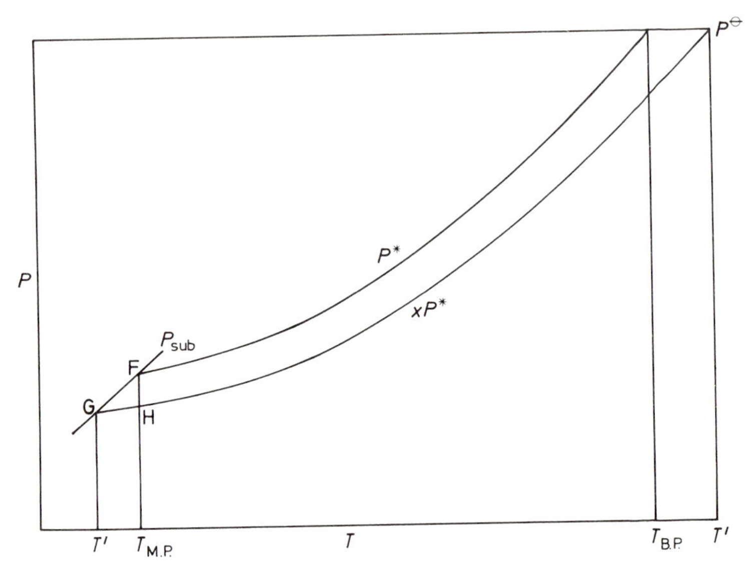

Figure 4.1 *Freezing point depression and boiling point elevation. (This drawing is purely schematic and not to scale.) For solutions obeying Raoult's law, $p = xp^*$, the b.p. is the temperature at which $p = P^{\ominus}$,* i.e., *where p crosses $P^{\ominus}$; similarly, the triple point (very close to the m.p.) has p equal to the sublimation pressure of ice. (The true m.p. refers to the solid at an imposed pressure, $P^{\ominus}$)*

An analogous treatment will give the depression of the freezing point of a solution, if the trivial effect of pressure on the condensed phases is ignored.§The freezing point of pure solvent and solvent in solution can then be identified with the intersections F and G, respectively, of the liquid and solid solvent v.p. curves.

By writing equation 4.6 for the sublimation curve GF and the liquid v.p. curve GH between the same two temperatures, the freezing points of the solution and pure solvent, and remembering that the vapour pressure at H, p^{H}, is given by Raoult's law as $p^{\mathrm{H}} = x_{\mathrm{A}}p^{\mathrm{F}}$ (where x_{A} refers to the solvent). Therefore

$$\ln\frac{x_{\mathrm{A}}p^{\mathrm{F}}}{p^{\mathrm{G}}} = -\frac{\Delta_{\mathrm{vap}}H}{R}\left(\frac{1}{T_0} - \frac{1}{T'}\right)$$

§*F* is actually the triple point and differs from the freezing point in the case of water by 0.01 K.

$$\ln\frac{p^{\mathrm{F}}}{p^{\mathrm{G}}} = -\frac{\Delta_{\mathrm{sub}}H}{R}\left(\frac{1}{T_0} - \frac{1}{T'}\right)$$

Subtracting these two equations then gives†

$$\ln x_{\mathrm{A}} = \frac{\Delta_{\mathrm{fus}}H_{\mathrm{A}}}{R}\left(\frac{1}{T_0} - \frac{1}{T'}\right) \tag{4.9}$$

Although equation 4.9 was derived for a non-volatile solute, it is equally true for volatile solutes, as will be shown in Chapter 8.

In both equations 4.8 and 4.9 further approximations are made for dilute solutions, giving a small ΔT, so that

$$(T_0^{-1} - T_1^{-1}) = (T_0 - T_1)/T_0T_1 \approx -\Delta T/T^2$$

Also

$$\ln x_{\mathrm{A}} = \ln(1 - x_{\mathrm{B}}) = -x_{\mathrm{B}} - x_{\mathrm{B}}^2/2 - x_{\mathrm{B}}^3/3 - \ldots$$
$$\approx -x_{\mathrm{B}} \approx -m_{\mathrm{B}}M_{\mathrm{A}}$$

where M_{A} is the molar mass in kg mol^{-1} and m_{B} is the *molality*‡ (*not* the mass!), *i.e.* moles of B in 1 kg of solvent.

With these substitutions

$$T_0 - T' = -\Delta T \approx x_{\mathrm{B}}RT_0^2/\Delta H_{\mathrm{A}} \approx km_{\mathrm{B}} \tag{4.10}$$

where $k = M_{\mathrm{A}}RT_0^2/\Delta H_{\mathrm{A}}$ is either the *freezing point depression* or *cryoscopic constant* if used with $\Delta_{\mathrm{fus}}H_{\mathrm{A}}$, or the *boiling point elevation* or *ebullioscopic constant* with $\Delta_{\mathrm{vap}}H_{\mathrm{A}}$, and can be used either for the determination of ΔT or, more often, the molar mass M_{B}.

In equation 4.9 x_{A} refers to the 'solvent' which freezes with the deposition of pure solid solvent in equilibrium with the solution, such as ice in equilibrium with a water–alcohol mixture. The situation is unchanged if ice is regarded as the solute and alcohol the solvent, when the solution can then be regarded as a saturated solution of ice in alcohol. The same equation with A and B interchanged can therefore be regarded as the *ideal solubility equation*, giving the ideal solubility in mole fractions.

†ΔH for sublimation must equal ΔH for fusion + evaporation.

‡Molality is used in thermodynamics, since it is independent of the density of the solvent and therefore of the temperature. It can also be established accurately by weighing and is readily converted into mole fractions. Molarity (mol dm^{-3}) has none of these advantages.

$$\ln x_B = \frac{\Delta_{fus} H_B}{R}\left(\frac{1}{T_0} - \frac{1}{T_1}\right) \tag{4.11}$$

Note that both ΔH and T_0 are characteristic of the solute and do not depend on the solvent at all. Nevertheless, equation 4.11 gives quite good approximations where there are no specific solvent–solute interactions. Equation 4.11 can be derived by a more rigorous method, and we shall return to solubility in Chapter 8.

EXAMPLES

4.1 The enthalpy change for the transition of orthorhombic to monoclinic sulphur at 95 °C is 11.3 kJ kg^{-1}, and the volume change is 1.4×10^{-5} m^3 kg^{-1}. What change in pressure (in bars) is required to change the transition temperature by 1 °C?

4.2 Calculate what additional pressure has to be applied to melt ice at −2 °C given the following data for 0 °C:

$$\Delta_f H_{ice} = -292.63 \text{ kJ mol}^{-1},\ \Delta_f H_{water} = -286.63 \text{ kJ mol}^{-1}$$

$$\rho_{ice} = 0.917 \text{ g cm}^{-3},\ \rho_{water} = 0.9993 \text{ g cm}^{-3}$$

Compare this pressure with the static pressure exerted by a skater of 70 kg on a skate of 4 mm × 25 mm contact area. (This can only be a rough assumption since skates are slightly curved.)

4.3 Estimate (state assumptions) the vapour pressure of mercury at 500 K if $\Delta_{vap} H = 58.6$ kJ mol^{-1} at its standard boiling point of 630 K.

4.4 Calculate the standard boiling point of water (at 1 bar), given that its normal b.p. is 373.15 K at 101.325 kPa and $\Delta_{vap} H$ is 40.9 kJ mol^{-1}.

4.5 The vapour pressure of liquids is frequently given by the equation

$$\ln p/\text{units} = A/T + B$$

(a) What is the thermodynamic significance of A and B and what are the underlying assumptions for an equation in this form, if A and B are to be independent of T?

(b) If the vapour pressure of water at 298 K is 3.167 kPa,

and 1 atm at 373 K, calculate A and B for (i) p/kPa and (ii) p/atm.
(c) What is the mean value of the enthalpy of vaporization?

4.6 Find the mean heat of vaporization of naphthalene from the two vapour pressures given:

$$p = 750 \text{ mmHg at } 217.368\ ^\circ\text{C}, \ p = 770 \text{ mmHg at } 218.536\ ^\circ\text{C}$$

4.7 From any relevant data given below, estimate the vapour pressure of $CHCl_3$ at 315 K
(a) making the usual assumptions, which should be stated; and
(b) making first-order correction for $\Delta_{vap}H$ varying with T.
For $CHCl_3$, $\Delta_{vap}H$ (normal b.p. 335 K) = 31.380 kJ mol^{-1}
For $CHCl_3$(l), $C_P(\text{l}) = 113.8$ J K^{-1} mol^{-1}
For $CHCl_3$(g), $C_P(\text{g}) = 65.8$ J K^{-1} mol^{-1}

4.8 The addition of 0.1405 g of cumene ($PhPr^i = C_9H_{12}$) to 11.360 g of palmitic acid ($C_{16}H_{32}O_2$) lowered the freezing point of the latter by 0.44 K from 335.4 K. Find $\Delta_{fus}H$ for palmitic acid.

4.9 'The boiling point elevation is independent of the nature of the solute.' Discuss the assumptions underlying this statement and, taking them to be correct, calculate the increase in the b.p. of water containing 1.0 mol% of urea. The enthalpy of vaporization of water is 40.8 kJ mol^{-1} at the boiling point.

ANSWERS

A4.1 Straight application of equation 4.3 gives

$$(\partial P/\partial T)_{fus} = \Delta H/T\Delta V = 11.3 \times 10^3/368 \times 1.4 \times 10^{-5}$$
$$= 21.9 \times 10^5 \text{ Pa K}^{-1}$$

No integration is required, and the answer is simply **21.9 bar**.

A4.2 From equation 4.3

$$(\mathrm{d}p/\mathrm{d}T)_{fus} = (\Delta H/T\Delta V)_{fus}$$

$$\Delta_{fus}H = \Delta_f H_{water} - \Delta_f H_{ice} = 6.00 \text{ kJ mol}^{-1}$$

$$\Delta_{fus}V = 18/0.9993 - 18/0.917 = 1.62 \text{ cm}^3 \text{ mol}^{-1}$$
$$= 1.62 \times 10^{-6} \text{ m}^3 \text{ mol}^{-1}$$

$$\therefore\ (\partial P/\partial T)_{\text{fus}} = 6 \times 10^3/(-1.62 \times 10^{-6} \times 273)$$
$$= 1.36 \times 10^7\ \text{Pa K}^{-1} = 136\ \text{bar K}^{-1}$$

The extra pressure to melt ice at −2 °C would therefore be **272 bar**.

The pressure exerted by a 70 kg man on an area of 100 mm^2 (10^{-4} m^2) is $9.81 \times 70 \times 10^4 = 687 \times 10^4$ Pa = 68.7 bar. This would seem to be sufficient only to melt ice at −0.5 °C; on the other hand, skating 'on edges' may well reduce the contact area sufficiently for a lubricating layer of water to be formed (friction may also help by heating the contact layer).

A4.3 Assuming $\Delta V \approx V_g$ and the vapour behaves like an ideal gas, and also neglecting the temperature variation of $\Delta_{\text{vap}}H$ between 500 and 630 K, we can use equation 4.6, giving

$$\ln p/\text{bar} = -(58\,600/8.314) \times (500^{-1} - 630^{-1}) = -2.91$$

$$\therefore\ p = 0.054\ \text{bar or } \mathbf{5400\ Pa}\ (cf.\ \text{observed value of 5250 Pa})$$

A4.4 Using the Clausius–Clapeyron equation (equation 4.4) with $\mathrm{d}\ln P = \mathrm{d}P/P$:

$$\mathrm{d}P/P\,\mathrm{d}T = 40\,900/8.314 \times (373.15)^2 = 0.035\,33\ \text{K}^{-1}$$

$\mathrm{d}P = -0.013\,25$ bar, $P = 1.0066$ bar (geometric mean of the pressures)

$$\therefore\ \mathrm{d}T = -0.013\,25/(1.0066 \times 0.035\,33) = -0.3726\ \text{K}$$

Therefore the standard boiling point of water is **372.78 K**.

A4.5 (a) From equation 4.6

$$\ln(p/P^{\ominus}) = \Delta_{\text{vap}}H/RT^{\ominus} - \Delta_{\text{vap}}H/RT$$

with $P^{\ominus}$ and $T^{\ominus}$ the standard pressure and boiling point (which for most published data would still be 1 atm rather than 1 bar). A is thus readily identified with $-\Delta_{\text{vap}}H/R$ (assumed independent of T), and if p is expressed as a fraction of $P^{\ominus}$:

$$B = \Delta_{\text{vap}}H/RT^{\ominus} = -A/T^{\ominus}$$

For any other units

$$\ln(p/P^{\ominus}) = \ln(p/\text{units}) - \ln(P^{\ominus}/\text{units})$$

so that

$$B = \Delta_{vap}H/RT^{\ominus} + \ln(P^{\ominus}/\text{units})$$
$$= -A/T^{\ominus} + \ln(P^{\ominus}/\text{units})$$

(b) Taking $P^{\ominus}$ as 101.325 kPa for a 373 K boiling point, equation 4.6 gives

$$\ln(3.167/101.325) = -A(298^{-1} - 373^{-1})$$
$$= -6.747 \times 10^{-4}A$$

$\therefore A =$ **−5136 K** for both (i) and (ii).

$B = 5136/373 =$ **13.77** for $P^{\ominus} = 1$ atm (or p in atm)

and $B = 13.77 + \ln 101.325 =$ **18.39** for p in kPa.

(c) $\Delta_{vap}H = -A \times R = 5136 \times 8.314$

$=$ **42700 J mol⁻¹** (mean value).

Note that $\partial\Delta_{vap}H/\partial T = \Delta C_P$ decreases with T since $C_P(\text{l}) > C_P(\text{g})$, but converge.

A4.6 $dp/p\,dT = \Delta_{vap}H/RT^2$

$$\therefore \Delta_{vap}H = RT^2\,dp/p\,dT$$
$$= 8.314 \times 491^2 \times 20/(760 \times 1.168)$$

$=$ **45.2 kJ mol⁻¹**

or, from the integrated equation

$$\Delta_{vap}H = -[RT_1T_2/(T_2 - T_1)]\ln(p_1/p_2)$$
$$= 8.314 \times 491^2 \times 0.026\,32/1.168$$

giving the same result. (*N.B.* The accuracy is limited by Δp.)

A4.7 (a) The principal assumptions are: $\Delta_{vap}H(T) \approx$ constant, $\Delta V \approx V_g \approx RT/p$ (since $V_g \gg V_l$), and b.p. is at 101.325 kPa. Equation 4.6 then gives

$$\ln p/\text{atm} = \Delta_{vap}H \times (T_0 - T)/RTT_0$$
$$= -31\,380 \times 20/8.314 \times 315 \times 335$$
$$= -0.7153$$

$\therefore p =$ **0.489 atm = 0.495 bar**

(b) $\Delta_{vap}H_T = \Delta_{vap}H_{335} + \int [C_P(g) - C_P(l)]\,dT$

$$= [31\,380 + (-48) \times (-20)] = 32\,340\ J\ mol^{-1}$$

But the mean value of $\Delta_{vap}H$ is then $31\,380 + 480 = 31\,860\ J\ mol^{-1}$. Substituting this value will give

$$\ln p/atm = -0.7262 = \ln 0.484$$

or $p =$ **0.490 bar**

For larger temperature differences it may be necessary to integrate equation 4.5 using $\Delta_{vap}H$ as a function of T. Using $\Delta_{vap}H_T = \Delta_{vap}H_{335} - [48\ J\ K^{-1}(T - 335\ K)]\ mol^{-1} = (47\,460 - 48T')\ J\ mol^{-1}$ (with $T' = T/K$)

$$d\ln p/dT = \Delta_{vap}H_T/RT^2$$

and after integration

$$\ln p/atm = [-47\,460 \times 20/(315 \times 335) - 48 \ln(315/335)]/8.314 = -0.7264$$

which is not significantly different from the previous result.

A4.8 Inverting equation 4.10:

$$\Delta_{fus}H = x_{cumene}RT^2/\Delta T$$

where T is the f.p. of pure palmitic acid.

$$x_{cumene} = 0.001\,17/(0.0444 + 0.00117) = 0.0260$$

$$\therefore \Delta_{fus}H = 0.0260 \times 8.314 \times 335.4^2/0.44 = \mathbf{54.6\ kJ\ mol^{-1}}$$

[Note that giving the result to one decimal place is of doubtful significance since the accuracy is limited by $\Delta T = 0.44$, and probably no better than 1%. This also justifies the use of the approximate formula equation 4.10, which neglects higher orders of x ($x^2/2 \approx 3 \times 10^{-4}$).]

A4.9 The principal assumption is that Raoult's law is obeyed, which means that there are no specific solvent–solute interactions. An approximate equation (equation 4.10) is therefore appropriate:

$$\Delta T = 0.01 \times 8.314 \times 373^2/40\,800 = \mathbf{0.28\ K}$$

Chapter 5

Open Systems

5.1 PARTIAL MOLAR QUANTITIES

Previous chapters have dealt with systems containing fixed amounts of substance at variable T and P. These are known as *closed systems*. If the amount of substance is to be varied, we speak of *open systems*. To describe such a system we need extra variables n_i with SI units of moles. For instance, the volume of a binary mixture will be a state function with variables T, P, n_1, and n_2, so that $V = V(T, P, n_1, n_2)$. V is the total volume for $n_1 + n_2$ moles and its total differential is

$$\mathrm{d}V = (\partial V/\partial T)\,\mathrm{d}T + (\partial V/\partial P)\,\mathrm{d}P + (\partial V/\partial n_1)\,\mathrm{d}n_1 + (\partial V/\partial n_2)\,\mathrm{d}n_2 \tag{5.1}$$

with the other three variables in each case being kept constant. Since T and P are normally kept constant, equation 5.1 will simply reduce to

$$\mathrm{d}V_{T,P} = (\partial V/\partial n_1)_{n_2}\,\mathrm{d}n_1 + (\partial V/\partial n_2)_{n_1}\,\mathrm{d}n_2 \tag{5.2}$$

The quantity V' (formerly $\bar{V}$) is defined as

$$V'_{\mathrm{B}} = (\partial V/\partial n_{\mathrm{B}})_{T,P,n_{\mathrm{A}}\ldots} \tag{5.3}$$

and is called the partial molar volume. It is illustrated in Figure 5.1. This kind of plot may be useful to determine V'_{B} from a plot of V for a solution containing $n_{\mathrm{B}} = m^{/}$ mol ($m^{/} = m/\mathrm{kg\ mol^{-1}}$) of solute in 1 kg of solvent, *viz.* a plot of V against molality, m (mol kg^{-1}). It is unsuitable if V' for both components are required, the 'method of intercepts' then being

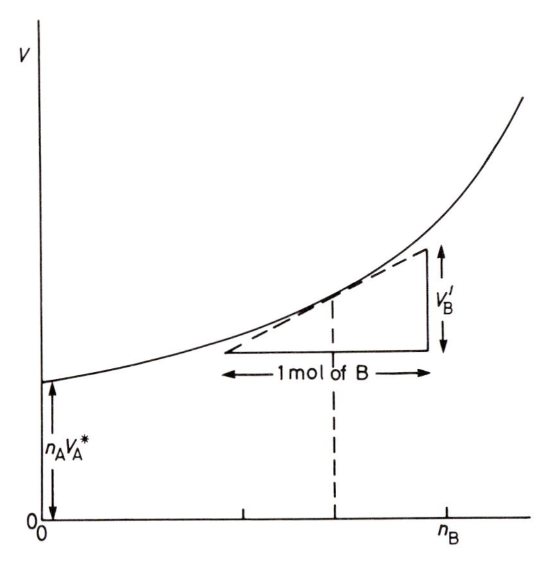

Figure 5.1 *Partial molar volume. In non-ideal solutions, volumes may not be additive, and the volume contribution per mole, V_B', may vary with concentration. V_A^* is the molar volume of pure A. If $n_A V_A^*$ corresponds to 1 kg of solvent, A, then n_B can be replaced by the molality, m*

appropriate. The latter will be discussed below (see equation 5.10).

Figure 5.1 clearly indicates that the partial molar volume of B varies with concentration, but would represent the increase in volume if 1 mole of B is added to a very large bulk of mixture, so as not to change the concentration. Since the last mole added is indistinguishable from the others in the mixture, V' is not only the increase per mole, but also the contribution per mole at that concentration, so that the total volume of the mixture, V_t, is

$$V_t = n_A V_A' + n_B V_B' + \dots \quad \text{(at constant } T \text{ and } P) \qquad (5.4)$$

Also a small increment in n_A, dn_A, would produce an increase $V_A' \, dn_A$, or generally

$$dV_t = V_A' \, dn_A + V_B' \, dn_B + \dots \quad \text{at } T, P \qquad (5.5)$$

But differentiation of each nV product in equation 5.4 would give $n \, dV + V \, dn$, so that

$$dV_t = n_A \, dV_A' + n_B \, dV_B' + \dots + V_A' \, dn_A + V_B' \, dn_B + \dots$$

which is only compatible with equation 5.5 if

$$n_A \, dV'_A + n_B \, dV'_B + \ldots = 0 \quad \text{at } T, P \tag{5.6}$$

Dividing both sides of equation 5.4 by Σn gives the volume for one mole of mixture and converts all n into mole fractions, x, so that

$$V_m = x_A V'_A + x_B V'_B + \ldots, \tag{5.7}$$

Similarly, equation 5.6 divided by Σn becomes

$$x_A \, dV'_A + x_B \, dV'_B + \ldots = 0 \quad \text{at } T, P \tag{5.8}$$

It follows from equations 5.7 and 5.8 that for a binary mixture

$$dV_m = V'_A \, dx_A + V'_B \, dx_B$$

$$\therefore \, dV_m/dx_B = V'_B - V'_A \tag{5.9}$$

$$(N.B. \; x_A + x_B = 1, \; \therefore \, dx_A/dx_B = -1)$$

which, used to eliminate V'_B from equation 5.7, gives

$$V_m = V'_A + x_B \, dV_m/dx_B \tag{5.10}$$

This is clearly the equation of a straight line ($y = c + mx$) with intercept V'_A and slope dV_m/dx_B for a plot of V_m *vs.* x_B, as shown in Figure 5.2. The line is, in fact, the tangent at any particular x_B. By equation 5.9 (or the symmetry in A and B) it follows that the intercept at $x_B = 1$ (or $x_A = 0$) must be V'_B.

This method of intercepts is used for the determination of all partial molar quantities. The measure of the amount need not be the mole. Thus if unit mass were to be used, the formal treatment would be exactly the same and only the symbols involved would change: V' would become the partial specific volume, v'; V_m would become the volume per unit mass, v; and x, the mole fraction, should then be replaced by w, the mass fraction. The tangent in Figure 5.2 would then make intercepts of the partial specific volumes at $w = 0$ and 1. These could easily be converted into partial molar quantities by dividing by moles/unit mass.

This extension of the method allows the direct determination of v' (and hence V') from data of densities of mixtures which are extensively tabulated, since the inverse of the density is the specific volume. It must be stressed that the two methods must not be mixed. Plots of molar volumes *vs.* mass fractions will give meaningless intercepts.

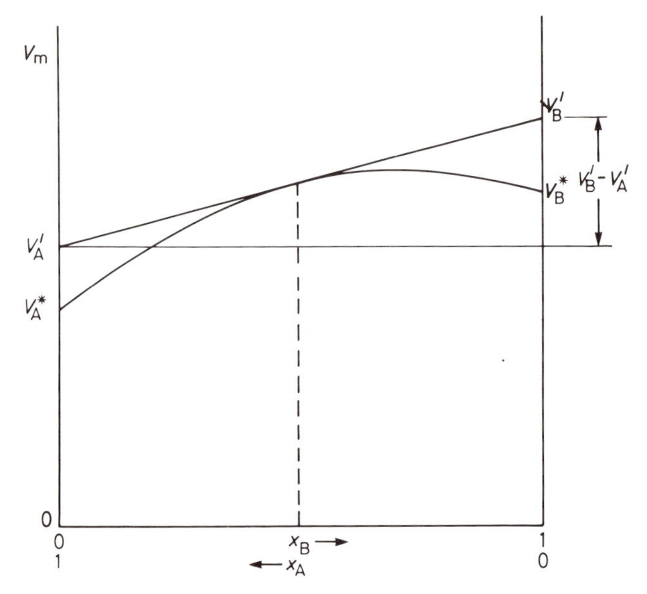

Figure 5.2 *The intercept method for partial molar volumes. If the volume per mole of mixture is plotted vs. mole fraction, the intercept at $x = 1$ of the tangent to the curve at any given concentration is the partial molar volume, V'. The intercept at $x = 0$ is V' for the other component*

While *intensive* properties, such as P, T, and ρ, are independent of the amount, *extensive* properties, such as U, H, G, *etc.*, are directly proportional to the amount, and these extensive functions can all give rise to partial molar quantities just as V does. Thus equations 5.1—5.11 all have their counterpart, with, for instance, H instead of V, and V' replaced by H'. Unlike the volume, however, the enthalpy is not directly measurable: only the change in enthalpy, ΔH, is, and we have to adapt the treatment accordingly.

5.2 HEAT OF MIXING

The process of mixing of A and B is understood to refer to the mixing of pure (denoted by *) components and can be written as a 'reaction':

$$n_A A + n_B B = n_A A(\text{soln}) + n_B B(\text{soln})$$

If the molar enthalpies of the pure components are H^* and the partial molar enthalpies in solution are H', then ΔH for mixing (products − reactants) gives

$$\Delta_{\text{mix}}H = n_{\text{A}}H'_{\text{A}} + n_{\text{B}}H'_{\text{B}} - n_{\text{A}}H^*_{\text{A}} - n_{\text{B}}H^*_{\text{B}}$$

which can be rearranged to

$$\Delta_{\text{mix}}H = n_{\text{A}}(H'_{\text{A}} - H^*_{\text{A}}) + n_{\text{B}}(H'_{\text{B}} - H^*_{\text{B}}) \qquad (5.11)$$

The quantities in brackets are known as *relative partial molar quantities*, but for the enthalpy of a solvent, A, $(H'_{\text{A}} - H^*_{\text{A}})$ is the *differential heat of dilution*, and for the solute, B, $(H'_{\text{B}} - H^*_{\text{B}})$ is the *differential heat of solution*. If $n_{\text{B}} = 1$ mole in equation 5.11, then $\Delta_{\text{mix}}H$ is the *integral heat of solution* for n_{A} moles of solvent, and is the figure most readily found in the literature for inorganic solutes.

The intercept method applied to unit amount of mixture will give the relative partial quantities per unit amount. Figure 5.3 shows the plot of the enthalpy of 1 mole of mixture *vs.* x_{B} and the intercepts of the tangents. These can be combined as in equation 5.7 to give

$$\Delta_{\text{mix}}H_{\text{m}} = x_{\text{A}}(H'_{\text{A}} - H^*_{\text{A}}) + x_{\text{B}}(H'_{\text{B}} - H^*_{\text{B}}) \qquad (5.12)$$

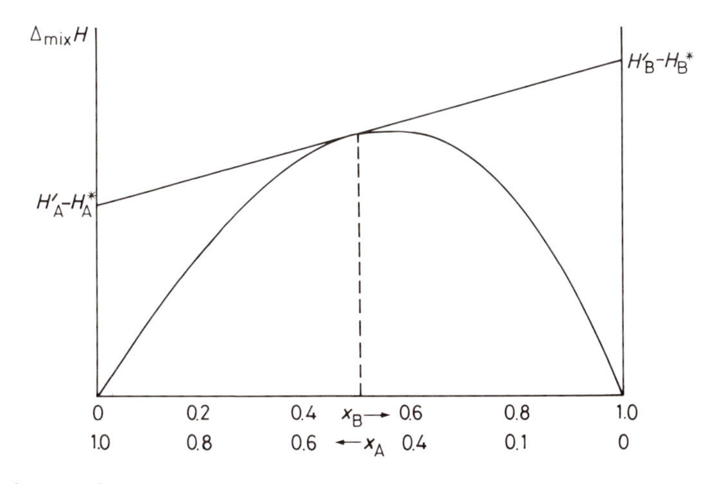

Figure 5.3 *The intercept method for relative partial molar enthalpies. $\Delta_{\text{mix}}H$ per mole of mixture is plotted vs. mole fraction. The intercepts of the tangent to the curve give the 'relative partial molar enthalpies' for the solvent ('differential heat of dilution') and solute ('differential heat of solution') by an equation analogous to equation 5.10*

5.3 THE CHEMICAL POTENTIAL

The partial molar Gibbs energy is given a special symbol, μ, and a special name, *chemical potential*, but by definition

$$\mu_A = G'_A = (\partial G/\partial n_A)_{T,P,n_B,\ldots} \quad (5.13)$$

or, by rearrangement

$$dG = \mu_A \, dn_A \quad \text{at } T, P, n_B$$

In general, there will be concentration changes in several components, or phases, or both, and the overall changes of Gibbs energy will be (at constant T and P)

$$dG = \mu_A \, dn_A + \mu_B \, dn_B \quad (5.14)$$

(*N.B.* Like all partial molar quantities, μ always refers to one mole of a substance.)

In the transfer of dn moles across a phase boundary from 1 to 2, *e.g.* from ice (1) to water (2) at or above the m.p., phase 1 loses dn and phase 2 gains dn moles, or $dn_2 = dn = -dn_1$. Thus $dG = (\mu_2 - \mu_1)\,dn$ and $\mu_1 > \mu_2$ for dG negative (or feasibility). It appears, therefore, that chemical substance will 'flow' from a higher to a lower chemical potential, and hence its name. If, on the other hand, $\mu_1 = \mu_2$, $dG = 0$ and there will be no net mass transfer. For equilibrium between phases 1 and 2 therefore:

$$\mu_{A,1} = \mu_{A,2}, \ \mu_{B,1} = \mu_{B,2}, \ \mu_{C,1} = \mu_{C,2} \ldots, \quad (5.15)$$

For an open system $G = G(T, P, n_A, n_B, \ldots)$, so equation 2.5 becomes

$$dG = (\partial G/\partial T)\,dT + (\partial G/\partial P)\,dP + (\partial G/\partial n_A)\,dn_A + (\partial G/\partial n_B)\,dn_B + \ldots \quad (5.16)$$

$$\text{At } T, P \qquad dG = \Sigma\, \mu\, dn \quad (5.17)$$

Observe that equation 5.17 corresponds to equation 5.5; likewise equation 5.6 becomes

$$n_A \, d\mu_A + n_B \, d\mu_B + \ldots = 0 \quad \text{at } T, P \quad (5.18)$$

and is known as the Gibbs–Duhem equation. It also follows from the definition of μ (equation 5.13) and equation A.1 (Appendix A) that

$$\partial\mu/\partial p = V' \ (\textit{cf.}\ \text{equation 2.7}) \quad (5.19)$$

$$\partial\mu/\partial T = -S' \text{ (\textit{cf}. equation 2.8)} \tag{5.20}$$

$$\partial(\mu/T)/\partial T = -H'T^{-2} \text{ (\textit{cf}. equation 2.10)} \tag{5.21}$$

$$\text{and } \Delta G = \Sigma\, n\mu \quad \text{(summed over products} - \text{reactants)} \tag{5.22}$$

Note that since $\mu = \mu(T, P, n_A, n_B, \ldots)$, the full version of equation 5.19 would be

$$(\partial\mu_i/\partial P)_{T,n_j} = V'_i$$

with corresponding changes for the subsequent two equations.

5.4 CHEMICAL POTENTIALS FOR IDEAL GASES AND IDEAL SOLUTIONS

From equation 2.7

$$\partial G/\partial P = V = RT/P$$

for 1 mole of an ideal gas, and it follows that

$$\mathrm{d}G_T = RT\,\mathrm{d}\ln P$$

or

$$G_B = RT\ln(P/P^\ominus) + k(T)$$

where $k(T)$ is an integration constant independent of P and equal to G_B when $P = P^\ominus$. It can also be shown that for an ideal gas at a partial pressure p, μ is independent of other gases present and equal to G_B.[8] Thus the chemical potential of an ideal gas of partial pressure p is

$$\mu = \mu^\ominus(T) + RT\ln(p/P^\ominus) \tag{5.23}$$

with the integration constant $k(T)$ written as $\mu^\ominus(T)$ since it equals μ at $p = P^\ominus$ [*i.e.* where $\ln(p/P^\ominus) = 0$].

The chemical potential of any component in a solution obeying Raoult's law can be easily found if its vapour is treated as an ideal gas:

For the solution $\quad \mu = \mu^\ominus(T) + RT\ln(xp^*/P^\ominus)$

For the pure solvent $\quad \mu^* = \mu^\ominus(T) + RT\ln(p^*/P^\ominus)$

By subtraction $\quad \mu = \mu^* + RT\ln x \quad (5.24)$

Equation 5.24 is often regarded as the definition of an ideal solution. It allows two other important properties to be deduced:

(a) Differentiation w.r.t. P gives

$$V' = V^*$$

$$\therefore \Delta_{\text{mix}} V = \Sigma n(V' - V^*) = 0$$

which shows that such solutions have no volume change on mixing.

(b) Differentiation of $\mu/T = \mu^*/T + R \ln x$ w.r.t. T gives

$$-H' T^{-2} = -H^* T^{-2}$$

$$\therefore \Delta_{\text{mix}} H = \Sigma n(H' - H^*) = 0 \tag{5.25}$$

which shows that such solutions have no enthalpy change on mixing.

The three characteristics of ideal mixtures or ideal solutions† can therefore be summarized as: obeying Raoult's law, no volume change on mixing, no heat change on mixing.

The Gibbs energy of ideal mixing is given by a formula analogous to equation 5.12 with μ replacing H':

$$\Delta_{\text{mix}} G_{\text{m}} = x_A(\mu_{\text{A}} - \mu_{\text{A}}^*) + x_{\text{B}}(\mu_{\text{B}} - \mu_{\text{B}}^*) = \Sigma\, xRT \ln x \tag{5.26}$$

Differentitation w.r.t. T (equation 2.8) gives

$$\Delta_{\text{mix}} S_{\text{m}} = -\Sigma\, xR \ln x \tag{5.27}$$

5.5 THE PHASE RULE

As shown in equation 5.15, for equilibrium between phases $dG = 0$, so that $\mu_{\text{A},1} = \mu_{\text{A},2}$. This must hold for all $\mathcal{P}$ phases, yielding $\mathcal{P} - 1$ constraints imposed by equilibrium for each component substance, thus reducing the number of variables which can be chosen compatible with equilibrium by $(\mathcal{P} - 1)\mathcal{C}$.

If there are $\mathcal{C}$ independent chemical components in each of $\mathcal{P}$ phases, $(\mathcal{C} - 1)\mathcal{P}$ concentration terms, together with two further variables, T and P, may at first be chosen.‡ The true number of independent variables, known as *degrees of freedom*, $\mathcal{F}$, are there-

†The difference between mixtures and solutions is sometimes said to be that solvent and solute are differentiated in solutions, but not in mixtures.

‡$\mathcal{C} - 1$, since the last concentration makes all fractions add up to unity.

fore $(\mathcal{C} - 1)\mathcal{P} - (\mathcal{P} - 1)\mathcal{C} + 2$

or

$$\mathcal{F} = \mathcal{C} - \mathcal{P} + 2 \tag{5.28}$$

This equation is known as the *phase rule*.

EXAMPLES

[In examples 5.1 to 5.3 it is convenient to introduce $v^{\ominus}$ as the specific volume equal to 1 $cm^3\,g^{-1}$, d as the relative density (taken to be 1 for a substance with density $= 1/v^{\ominus}$), $m' = m/m^{\ominus}$, and $m^{\ominus} = 1$ mol kg^{-1}(solvent). The following molar masses may be used for the examples. $M/$g mol^{-1}: H_2O, 18; H_2O_2, 34; HF, 20; benzene, 78; toluene, 92]

5.1 The volume of a dilute solution of KCl of molality m (containing m' moles of KCl in 1 kg of water, with $m' = m/$mol kg^{-1}) is given by the equation $V/cm^3 = 1003 + 27.15m' + 1.744(m')^2$. Obtain equations for V'(KCl) and $V'(H_2O)$ and evaluate them for $m' = 0$ and $m' = 0.5$ (*N.B.* $m' = 0$ means infinite dilution).

5.2 Using the data below for wt% HF and relative density, d, find the partial molar volumes of H_2O in solutions containing 75% and 90% HF.

Wt% HF	0	50	60	70	75	80	85	90	95	100
d	1.000	1.198	1.235	1.258	1.262	1.259	1.240	1.178	1.089	1.000

5.3 The relative density, d, of H_2O_2–H_2O mixtures is given by $d = a + bw + cw^2$, where w is the weight fraction of H_2O_2 and $a = 0.9612$, $b = 0.2765$, $c = 0.1196$ for the mixtures at 96 °C. Show that the partial specific volumes of water and hydrogen peroxide are given by

$$v'(H_2O)/v^{\ominus} = d^{-2}(a + 2bw + 3cw^2)$$

and

$$v'(H_2O_2)/v^{\ominus} = d^{-2}(a + 2bw + 3cw^2 - b - 2cw)$$

Hence find the partial molar volumes of H_2O_2 and H_2O in a mixture containing 10 wt% H_2O_2.

5.4 (a) Noting the approximately linear portion between 0.68 and 0.83 in the following data for the system MeOH(x)–PhCl($1 - x$), find the relative partial molar enthalpies for

MeOH and PhCl at $x = 0.75$, and use the results to calculate the heat absorbed on mixing 3 moles of MeOH and 1 mole of PhCl. Check this value with the given $\Delta_{mix}H$ per mole.

$10^4\ x$	750	2187	5875	6848	7220	7512	7869	8277	9574
$\Delta_{mix}H/\mathrm{J}$	507.6	631.7	402.5	307.2	267.2	236.5	198.3	155.9	32.1

(b) Evaluate graphically the relative partial molar enthalpies of benzene and toluene at $x_b = 0.6$ from the data given below, and use the results to calculate the heat absorbed on mixing 6 moles of benzene and 4 moles of toluene, with $M_r = 78$ and 92, respectively. Check this value with the given $\Delta_{mix}H$ per kg of solution.

Wt% benzene	10	20	30	40	50	60	70	80	90
$\Delta_{mix}H/\mathrm{J\ kg^{-1}}$	334	602	828	916	933	920	778	519	263

5.5 For a wide range of binary liquid mixtures the heat of mixing is well represented by an equation of the form $\Delta H = ax^3 - (a + b)x^2 + bx$, where a and b are constants, and x is the mole fraction of the solute. Show that this expression leads to the expected value of the differential heat of solution when $x = 1$, and to $H' - H^* = b$ for $x = 0$.

5.6 The heat of mixing of acetone (2) and chloroform (1) may be described in terms of the equation

$$\Delta_{mix}H/\mathrm{J\ mol^{-1}} = 7640x_1x_2 + 2160x_1x_2(x_1 - x_2)$$

Find the heat of solution of 1 mole of acetone in a large (assume infinite) amount of acetone–chloroform mixture containing 10 mol% acetone.

5.7 Give an estimate of the minimum amount of work required to separate a volume of air (assumed to consist of 1 mole of oxygen and 4 moles of nitrogen) at 20 °C and 1 atm into its constituents, each at 20 °C and 1 atm.

5.8 Show that the 'triple point' of any single substance will be at a fixed temperature and pressure.

5.9 Using the phase rule, explain what equilibrium vapour pressure and concentration changes will occur at constant temperature as increasing amounts of water are added to a non-volatile salt, X, which forms a single hydrate, $X{\cdot}H_2O$, and then dissolves in water.

5.10 If the chemical potential of A is given by the expression

$$\mu_A - \mu_A^* = RT\ln x_A + Bx_B^2$$

what will be the expression for $\mu_B - \mu_B^*$? [See the Gibbs–Duhem equation.]

ANSWERS

A5.1 By definition $V'_B = (\partial V/\partial n_B)$ at constant n_A (equation 5.3), but, with A = H_2O, n_A in 1 kg of A is 1 kg/18 g mol^{-1} = 55.5 mol (and therefore constant), and n_B equals $m^/$ mol (in this case, B = KCl).

$$\therefore\ V'(\text{KCl})/\text{cm}^3\ \text{mol}^{-1} = 27.15 + 2 \times 1.744m^/$$

$V'(H_2O)$ is obtained from equation 5.7 as

$$[V_t/\text{cm}^3 - (27.15 + 3.488m^/)m^/]/n_A$$

so that for H_2O

$$V'/\text{cm}^3\ \text{mol}^{-1} = [1003 - 1.744(m^/)^2]/55.5$$

∴ for KCl

$$V'/\text{cm}^3\ \text{mol}^{-1} = \mathbf{27.15}\ \text{at}\ m^/ = 0$$

$$\text{and}\ \mathbf{28.89}\ \text{at}\ m^/ = 0.5$$

and for H_2O

$$V'/\text{cm}^3\ \text{mol}^{-1} = \mathbf{18.07}\ \text{at}\ m^/ = 0$$

$$\text{and}\ \mathbf{18.06}\ \text{at}\ m^/ = 0.5.$$

A5.2 Using the intercept method shown in Figure 5.2, but for unit mass and mass-fraction instead of 1 mole of mixture and mole fraction, a plot of $v/v^{\ominus} = 1/d$ *vs.* wt% HF is prepared and tangents are drawn at 75 and 90%. The intercepts give the partial specific volumes, $v' = V'/M$, but are difficult to determine accurately, since they involve extrapolation. It is probably better therefore to calculate the intercepts from the slope, m, of the chord through the neighbouring points and then use m to calculate the intercept, c, from the tangent $y' = mx' + c$ through the point (x', y'). If $y = 1/d$ and x = wt%, the required points are:

Wt%	70	75	80	85	90	95
$1/d$	0.7949	0.7924	0.7943	0.8065	0.8489	0.9183

At 75% this gives

$$c = v'/v^{\ominus} + 0.7924 + 6 \times 10^{-3} \times 0.75 = 0.797$$

$$\therefore v' = 0.797 \text{ cm}^3 \text{ mol}^{-1}$$

and

$$V' = \mathbf{14.35\ cm^3\ mol^{-1}}$$

At 90%

$$v'/v^{\ominus} = 0.8489 - 1.118 \times 0.90 = -0.1575$$

or

$$V' = \mathbf{-2.83\ cm^3\ mol^{-1}}$$

Note that negative partial molar volumes are perfectly feasible and arise from a contraction in volume on addition of one component. It can best be understood by considering the contraction which occurs on addition to toluene to a volume of expanded polystyrene, and incidentally shows that the negative v' is due to the structural collapse of the *other* component.

A5.3 For the equation $d = a + bw + cw^2$ the slope of the tangent of $1/d$ *vs.* w is

$$\partial(d^{-1})/\partial w = -d^{-2}\partial d/\partial w = -d^{-2}(b + 2cw)$$

From the equation for a straight line the intercept at $w = 0$, is

$$v'(H_2O)/v^{\ominus}$$
$$= d^{-1} + d^{-2}(b + 2cw)w = (a + 2bw + 3cw^2)/d^2$$

By equation 5.9

$$v'(H_2O_2)/v^{\ominus} = \partial v/\partial w + v'(H_2O)/v^{\ominus}$$
$$= (a + 2bw + 3cw^2 - b - 2cw)/d^2$$

Inserting the given values:

$$d = 0.99,\ v'(H_2O)/v^{\ominus} = 1.04,\ v'(H_2O_2)/v^{\ominus} = 0.73$$

$$V'(H_2O) = \mathbf{18.7\ cm^3\ mol^{-1}}$$

and

$$V' \, (H_2O_2) = \mathbf{24.8\ cm^3\ mol^{-1}}$$

A5.4 (a) The best straight line through the given points in the given range has intercepts

$$H' - H^* = \mathbf{1032\ J\ mol^{-1}} \text{ at } x = 0 \text{ (for PhCl)}$$

and

$$\mathbf{-27.0\ J\ mol^{-1}} \text{ at } x = 1 \text{ (for MeOH)}$$

The heat of mixing 3 moles of MeOH and 1 mole of PhCl is therefore $\Delta_{mix}H = 3(-27) + 1032 = \mathbf{951\ J}$, or 237.7 J/mol of mixture. This is in complete agreement with the interpolated value at $x = 0.75$.

(b) Note warning in text concerning mixed plots – either ΔH *vs.* x or Δh *vs.* w must be used. The latter is simpler and only requires the tangents to be drawn at w corresponding to $x_b = 0.6$, readily calculated from $w_b = x_b M_b/(x_b M_b + x_t M_t)$, giving $w_b(0.6) = 0.560$.

The tangent at this point gives intercepts 1020 J kg^{-1} at $w = 0$ (toluene) and 860 J kg^{-1} at $w = 1$ (benzene). The corresponding molar quantities are

$$H'_t - H^*_t = 0.092 \times 1020 = 93.8\ \text{J mol}^{-1}$$

and

$$H'_b - H^*_b = 0.078 \times 860 = 67\ \text{J mol}^{-1}$$

For 6 moles (368 g) of benzene and 4 moles (468 g) of toluene (*i.e.* a total of 836 g)

$$\Delta H = 0.468 \times 1020 + 0.368 \times 860 = \mathbf{778\ J}$$

giving $778/0.836 = 931$ J kg^{-1}, which corresponds to the value at $w = 0.56$ from the graph.

A5.5 Differentiation of ΔH:

$$\partial \Delta H/\partial x = 3ax^2 - 2(a + b)x + b$$

gives the slope of the tangent at x, which becomes b at $x = 0$. The equation of the tangent at $x = 0$ is clearly $y = xb$ with an intercept at $x = 1$, $H' - H^* = b$. This can also be obtained from the general expression, as follows.

Elimination of V'_A from equations 5.9 and 5.10 gives

$$V'_B = V_m + (1 - x_B)\partial V_m/\partial x_B$$

or the corresponding equation

$$H' - H^* = \Delta H + (1 - x)\partial \Delta H/\partial x$$

with $\Delta H = 0$ both at $x = 0$ or $x = 1$, and the second term equal to b for $x = 0$ and 0 for $x = 1$. This latter result is to be expected since it represents the increase in enthalpy when one mole of solute is added to an infinite bulk of pure solute (*i.e.* 0).

A5.6 Again, application of equation 5.10 to $\Delta_{mix}H$ gives as one form

$$H'_2 - H^*_2 = \Delta_{mix}H - x_1\partial\Delta_{mix}H/\partial x_1$$

There are several ways of obtaining the differential coefficient:
(a) replacing x_2 by $1 - x_1$ (trivial, but lengthy!)
(b) remembering $\partial x_2/\partial x_1 = -1$, or
(c) numerically .
(b) gives

$$a(x_1 - x_2) - b(x_1 - x_2)^2 + 2bx_1x_2 \rightarrow 7106\ \mathrm{J\ mol^{-1}}$$

or an overall expression

$$H'_2 - H^*_2 = ax_1^2 + bx_1[(x_1 - x_2)^2 - x_2]$$

(c) gives $\Delta H/\mathrm{J\ mol^{-1}} = 843.1200$ at $x_1 = 0.9$ and 842.4093 at $x_1 = 0.9001$ with a slope of $-7107\ \mathrm{J\ mol^{-1}}$ and, by substitution, the result

$$H'_2 - H^*_2 = 843 - 0.9 \times (-7107) \approx \mathbf{7240\ J\ mol^{-1}}$$

A5.7 The minimum energy of separation is $-\Delta_{mix}G = -\Sigma\, xRT\ln x$ per mole of ideal mixture (see equation 5.26). For 5 moles

$$\Delta_{mix}G = RT(\ln 0.2 + 4\ln 0.8) = -6095\ \mathrm{J}$$

The minimum energy required for unmixing is therefore **6095 J**.

A5.8 From the phase rule (equation 5.28) $\mathcal{F} = \mathcal{C} - \mathcal{P} + 2$. At the triple point there are three phases in equilibrium (l, g, s), but only one component. $\therefore\ \mathcal{F} = 1 - 3 + 2 = 0$, so that no parameters can be chosen. They are unique. This can also be seen by counting the initial variables, T, P, but no concentration terms, and subtracting the number of constraints introduced by the equilibrium conditions, $\mu_g = \mu_l = \mu_s$, *i.e.* $2 - 2 = 0$.

A5.9 Initially, all the water added will be taken up to form the hydrate, $X{\cdot}H_2O$. Throughout, there will be two components, X and H_2O, and during this early stage, since solids do not normally mix (*i.e.* form solutions), there will be the three phases X, $H_2O(g)$, and $X{\cdot}H_2O$.

According to the phase rule, $\mathcal{F} = \mathcal{C} - \mathcal{P} + 2 = 1$. This one degree of freedom is taken up by the choice of T, and therefore p is fixed. When all X is converted into $X{\cdot}H_2O$, addition of more water will start to dissolve $X{\cdot}H_2O$. At this stage there will still be three phases [$X{\cdot}H_2O$, solution, $H_2O(g)$], $\mathcal{F} = 1$ and p is again fixed, albeit at a different value (corresponding to the saturated solution formed).

When all $X{\cdot}H_2O$ is dissolved, the solid phases will both have disappeared, leaving only two phases [solution and $H_2O(g)$]. Now $\mathcal{F} = 2$ and p can therefore vary at constant T. As an alternative, one can regard the mole fraction, x, as the variable and p determined by the value of x, but only one of them can be chosen as independent variable.

In the earlier stages x is fixed as 1 for all solids, and $x = x_{sat}$ in the solution formed in the second stage.

A5.10 By equation 5.18, $\Sigma\, x\, \mathrm{d}\mu = 0$ at constant T and P. Remembering that μ_A^* does not vary with x, and $x_B = 1 - x_A$

$$\mathrm{d}\mu_A = RT\,\mathrm{d}x_A/x_A + 2Bx_B\,\mathrm{d}x_B$$

$$\therefore\ \mathrm{d}\mu_B = -x_A\,\mathrm{d}\mu_A/x_B = -RT\,\mathrm{d}x_A/x_B + 2Bx_A\,\mathrm{d}x_A$$

$$(\mathrm{d}x_A = -\mathrm{d}x_B)$$

so that on integration we get

$$\mu_B = RT\ln x_B + Bx_A^2 + I$$

where the integration constant I must make $\mu_B = \mu_B^*$ for $x_B = 1$. Since the rest of the expression vanishes for $x_B = 0$, $I = \mu_A^*$, giving

$$\mu_B = \mu_B^* + \boldsymbol{RT\ln x_B} + \boldsymbol{Bx_A^2}$$

Chapter 6

Non-Ideal Gases

6.1 EQUATIONS OF STATE

Relationships between the variables P, V, and T are known as *equations of state*. In the case of non-ideal gases, several such equations exist. The van der Waals equation takes into account the effective volume occupied by the gas molecules by a term b and the forces of attraction between the molecules by a/V^2 to give

$$(P + an^2/V^2)(V - nb) = nRT$$

or for 1 mole

$$(P + a/V^2)(V - b) = RT \tag{6.1}$$

By neglecting second-order terms, this can be reduced to

$$PV = RT + BP \tag{6.2}$$

with $B = b - a/RT$

With further terms added, equation 6.2 becomes a more general series known as the *virial equation*:

$$PV = RT + BP + CP^2 + DP^3 + \dots$$

with the virial coefficients $B, C, D, \dots$ all dependent on T.

Finally, a commonly used equation, introducing a compression factor† Z, is

$$PV = Z(T, P)RT \tag{6.3}$$

6.2 CHEMICAL POTENTIAL AND FUGACITIES

The chemical potential for an ideal gas is (equation 5.23)

$$\mu = \mu^{\ominus}(T) + RT\ln(p/P^{\ominus}) \tag{6.4}$$

†This used to be termed 'compressibility factor'.

At low pressure, all non-ideal gases approach the ideal state and their equations of state tend to converge to $PV = RT$. This is used in the expression for the chemical potential of a non-ideal gas, which is

$$\mu = \mu^{\ominus}(T) + RT\ln(f/P^{\ominus}) \tag{6.5}$$

together with

$$\lim_{p \to 0}(f/p) = 1$$

where the *fugacity*, f, the 'thermodynamic' pressure, is a function both of T and P, with the dimensions of pressure.

The limiting condition defines the standard chemical potential as that of the ideal gas – which the real gas approaches at low pressure – extrapolated to $P^{\ominus}$. This has the important implication that $\mu^{\ominus}$ is still only a function of T and all non-ideal behaviour is taken account of by f, which therefore varies both with T and P. For a non-ideal gas, therefore, f need only equal P at $P = 0$ and may depart from P as the pressure increases from 0 to $P^{\ominus}$. Hence $\mu^{\ominus}$ refers to the 'extrapolated' value at $P^{\ominus}$, though the departure at $P^{\ominus}$ is usually negligible.

Taking μ for a pure (*), non-ideal gas, and subtracting μ of the equivalent ideal gas from it (equation 6.5 minus equation 6.4, but with $p = P$ for a single gas):

$$\mu^* - \mu^{id} = RT\ln(f^*/P) = RT\ln\gamma \tag{6.6}$$

where γ $(= f^*/P)$ is the *fugacity coefficient*, a correction factor to bring the pressure to the 'thermodynamic pressure', f, for a real gas. Differentiating equation 6.6 w.r.t. P and dividing by RT gives

$$\begin{aligned}(\partial/\partial P)\ln\gamma &= (\partial\mu/\partial P - \partial\mu^{id}/\partial P)/RT \\ &= (V' - V^{id})/RT = V'/RT - 1/P\end{aligned}$$

or

$$\ln\gamma = \int_0^P (V'/RT - 1/P)\,dP \tag{6.7}$$

For a pure gas $V' = V$, and any of the expressions 6.1—6.3 may be used to substitute for V'/RT in equation 6.6. For example:

From equation 6.2 $\qquad \ln\gamma = BP/RT$

From equation 6.3 $\qquad \ln\gamma = \int_0^P (Z - 1)\,d\ln P \qquad (6.8)$

Equation 6.7 provides a useful graphical method for the determination of the fugacity coefficient,[7] since the integral is clearly the area on a plot of Z *vs.* $\ln P$ bounded by the lines $Z = 1$ (ideal gas), Z, and $\ln P$ (see Figure 6.1). The lower limit $P = 0$, which makes $\ln P = -\infty$, is, of course, unattainable, but in most cases the area between 0 and $P^{\ominus}$ is negligible. (It can be estimated analytically; see examples.)

The actual plot makes use of the *principle of corresponding states*, according to which the behaviour of all gases can be closely matched in terms of their *reduced variables:* P_R $(= P/P_c)$, V_R $(= V/V_c)$, and T_R $(= T/T_c)$, where P_c, V_c, and T_c are the *critical constants*.

In the log plot used, the shift of origin and upper limit of $\ln P$ by $\ln P_c$ leaves the value of the integral unchanged.

It is worth noting that for $T_R < 2$ the initial area will be negative, so that for low T_R and $P_R <$ *ca.* 10, γ will be less than 1. Values of $\gamma > 1$ are only found for gases with $T_R > 2$ at high pressures.

6.3 GASES IN IDEAL MIXTURES

Except in the vicinity of the critical state, gas molecules are widely separated. It is therefore reasonable to expect that gas mixtures will approximate to ideal mixtures. Using y for the mole fraction of a gas, and remembering that μ^* for a mixture is a function of both T and P (see equation 5.19), equation 5.24 becomes

$$\mu = \mu^*(T, P) + RT \ln y \tag{6.9}$$

Equating equation 6.9 to equation 6.5, since both are valid expressions for the same gas, gives

$$\mu^*(T, P) + RT \ln y = \mu^{\ominus}(T) + RT \ln (f/P^{\ominus})$$

or

$$RT \ln (f/P^{\ominus}) = RT \ln y + \mu^*(T, P) - \mu^{\ominus}(T)$$

But by letting $y \rightarrow 1$ at constant T and *total* pressure, P, $f \rightarrow f^*(T, P)$ and

$$RT \ln (f^*/P^{\ominus}) = \mu^*(T, P) - \mu^{\ominus}(T)$$

so that

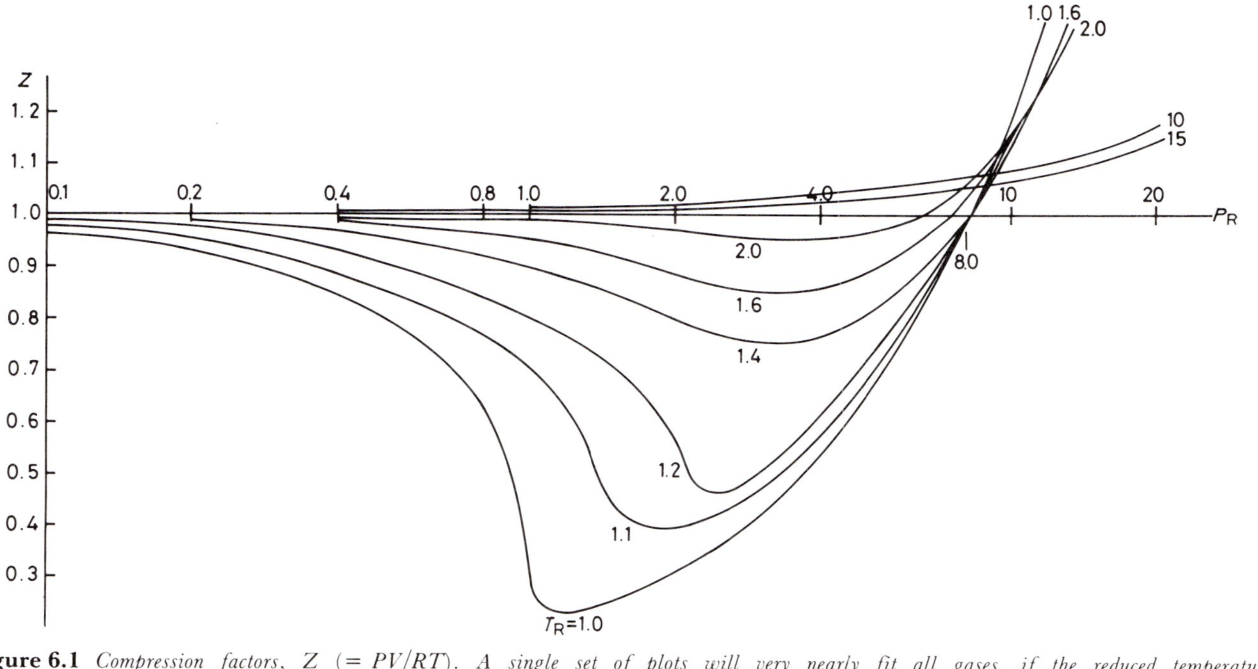

Figure 6.1 *Compression factors, Z (= PV/RT). A single set of plots will very nearly fit all gases, if the reduced temperature* ($T_R = T/T_c$) *and the reduced pressure* ($P_R = P/P_c$) *are taken as the variables.* [$T_c \approx 1.6T_{bp}$, $P_c \approx 3$—9 MPa *(notable exceptions are* H_2 *and* H_2O*).*] *Note that most* $Z < 1$ *up to about* $P_R = 8$*, so that* $\ln \gamma$ *would then be negative or* $\gamma < 1$ *(see equation 6.8) (Adapted from ref. 7)*

$$RT\ln(f/P^{\ominus}) = RT\ln y + RT\ln[f^*(T, P)/P^{\ominus}]$$

or

$$f = yf^*(T, P) \tag{6.10}$$

This important result is known as the Lewis and Randall rule, which can also be written in terms of fugacity coefficients by substituting p/P for y into equation 6.10 and rearranging, so that

$$\gamma = f/p = f^*/P \tag{6.11}$$

As will be shown in Chapter 8 (see equation 8.2), the true equilibrium constant, $K^{\ominus}$, is composed of the variable terms in the chemical potential expression. In the case of a non-ideal gas equilibrium, $f = \gamma p$ takes the place of p, so that K_f, and not K_P, is the equilibrium constant independent of the total pressure. Equation 3.5 has to be rewritten as

$$K^{\ominus}(P)^{\Delta v} = K_f = K_P K_\gamma = K_\gamma K_x P^{\Delta v} \tag{6.12}$$

and similarly, f has to replace p in equations 3.3 and 3.4.

None of these changes are important for reactions at atmospheric pressure, but could produce appreciable corrections for high-pressure reactions, *e.g.* when high T_R gases combine to form a low T_R product.

6.4 ADIABATIC FLOW PROCESSES

As shown in equation 2.16

$$(\partial H/\partial P)_T = V - T(\partial V/\partial T)_P \tag{6.13}$$

Substituting $V = RT/P$ for an ideal gas, $(\partial H/\partial P)_T = 0$, but in terms of the van der Waals constants of equation 6.2:

$$(\partial H/\partial P)_T = B - T\partial B/\partial T = b - 2a/RT \tag{6.14}$$

This allows enthalpy and C_P variations with pressure to be computed. Since $C_P = (\partial H/\partial T)_P$:

$$(\partial C_P/\partial P)_T = (\partial/\partial T)(\partial H/\partial P)_T = 2a/RT^2 \tag{6.15}$$

When a non-ideal gas streams through a small orifice or porous plug, it undergoes a change in temperature and pressure. When this process is carried out adiabatically, the resulting $\mathrm{d}T/\mathrm{d}P$ is the Joule–Thomson coefficient, μ_{JT}. Considering one mole of gas being pushed through the orifice, the work done by the gas is

$-P_1\int_0^V \mathrm{d}V$ before and $P_2\int_0^V \mathrm{d}V$ after the orifice. Therefore, the net work done by the gas, ignoring any losses due to turbulence or changes in kinetic energy of the gas, which are small, is

$$w = P_2V_2 - P_1V_1 \tag{6.16}$$

Since we are considering an adiabatic process, the First Law expression $q = w + \Delta U = 0$ applies, and this combined with equation 6.16 becomes

$$P_2V_2 + U_2 = P_1V_1 + U_1$$

or

$$H_2 = H_1$$

The process is therefore at constant H (*i.e.* is *isenthalpic*), so that

$$\begin{aligned}\mu_{\text{JT}} &= (\partial T/\partial P)_H = -(\partial H/\partial P)_T/(\partial H/\partial T)_P \\ &= -[V - T(\partial V/\partial T)_P]/C_P \end{aligned} \tag{6.17}$$

and for a van der Waals gas

$$\mu_{\text{JT}} = (2a/RT - b)/C_P \tag{6.18}$$

The sign of μ_{JT} will change when $2a/RT = b$, *i.e.* when $T = T_{\text{inv}} = 2a/Rb$ (the inversion temperature). Gases will cool below that temperature, and this is utilized in the Linde process for their liquefaction. For He and H_2 the inversion temperature lies well below room temperature, so they have to be pre-cooled before the Linde process can be used.

The Joule–Thomson treatment can be generalized to other adiabatic steady-flow processes in which the orifice may be replaced or supplemented by turbines or compressors; the additional work (by the gas) is w_s (for 'shaft work'), so that equation 6.16 becomes

$$w = P_2V_2 - P_1V_1 + w_s \tag{6.19}$$

and in these general adiabatic flow processes

$$\Delta H + w_s = 0 \tag{6.20}$$

EXAMPLES

(Unless otherwise stated, $P^{\ominus} = 1$ bar $= 10^5$ Pa.)

6.1 Show that the van der Waals equation, equation 6.1, may be

reduced to equation 6.2 by neglecting second-order terms and use it to
(a) express the second virial coefficient, B, in terms of the van der Waals coefficients, and
(b) evaluate B for CO_2 at 60 °C in $cm^3\,mol^{-1}$.
[For CO_2 at 60 °C: $a = 3.1 \times 10^6\ cm^6\,atm\,mol^{-2}$, and $b = 34.0\ cm^3\,mol^{-1}$.]

6.2 Assuming the equation $PV = A + BP$ for CH_4 at 0 °C, with $A = 22\,410\ cm^3\,atm\,mol^{-1}$ and $B = -55.6\ cm^3\,mol^{-1}$, calculate the change in chemical potential when one mole of CH_4 is isothermally compressed from 1 to 80 atm.

6.3 By suitable graphical integration, estimate f and γ for CH_4 at −70 °C and at 10 and 50 atm respectively from the following data:

P/atm	1	10	20	30	40	50
PV/RT	0.994	0.940	0.870	0.793	0.703	0.594

6.4 The compression factor, Z, for CO_2 at 0 °C and $P^{\ominus} = 1$ atm is given by

$$Z = 1 - 2.38 \times 10^{-3} P' + 5.22 \times 10^{-6} (P')^2$$

(with $P' = P/P^{\ominus}$)
Calculate the fugacity for CO_2 at 0 °C and 100 atm.

6.5 Assuming that for N_2, $\ln \gamma = 4.1 \times 10^{-4} p/P^{\ominus}$, calculate γ and f at a total pressure of 15 MPa for nitrogen in
(a) pure nitrogen, and
(b) 3 moles of an ideal mixture containing 1 mole N_2.

6.6 If the only reaction occurring in a reactor at 500 K and 200 atm is $CO + 2H_2 = CH_3OH(g)$, with $\Delta G^{\ominus} = 19.08\ kJ\,mol^{-1}$ ($P^{\ominus} = 1$ atm), estimate the per cent CO converted into CH_3OH under these conditions from a 1 : 2 mixture of CO and H_2, both assumed to behave as ideal gases, if CH_3OH has a compression factor $Z = 1.00 - 4 \times 10^{-3} P/\text{atm}$.

6.7 Using the truncated van der Waals equation $V = RT/P + b - a/RT$ with the data given below, cal ulate for NH_3 at 750 K:

(a) the changes in C_P and H due to an increase from 1 to 40 atm,
(b) the Joule–Thomson coefficient at 40 atm, and
(c) the inversion temperature of NH_3.

For NH_3:
$$a = 4 \times 10^6 \text{ cm}^6 \text{ atm mol}^{-2}$$
$$b = 36 \text{ cm}^3 \text{ mol}^{-1}$$
$$C_P^{\ominus} = 48.1 \text{ J K}^{-1} \text{ mol}^{-1}$$

6.8 With the above data, calculate the fugacity of NH_3 at 100 atm and 750 K.

6.9 (a) In the synthesis of ammonia from H_2 and N_2 at 723 K and 600 atm total pressure, the equilibrium mixture contained 11.6 mol% N_2 and 53.6 mol% NH_3. Calculate K_P and compare it with the value obtained from $\Delta_f G_{723} = 29.83 \text{ kJ mol}^{-1}$.
(b) Show that the equation $PV = RT + BP + CP^2 + DP^3$ leads to
$$\log(f/P) = \log \gamma = aP + bP^2 + cP^3$$
(c) Assuming cP^3 can be neglected in the expression for $\log \gamma$, find the appropriate correction factor to K_P from the data below, and comment on the result.

	$10^4 a/\text{atm}^{-1}$	$10^8 b/\text{atm}^{-2}$
NH_3	−0.766	−3.215
N_2	1.774	7.770
H_2	1.229	4.527

ANSWERS

A6.1 (a) In $PV - Pb + a/V - ab/V^2 = RT$, all terms containing a or b are correction terms, with the second-order term containing ab being neglected except at very high pressures. Also, the first-order correction term $a/V = aP/PV \approx aP/RT$. With these approximations:
$$PV = RT + P(b - a/RT) = A + BP$$
and $\boldsymbol{B = b - a/RT}$
(b) Note that b and a/RT have to be in the same units (use Appendix B).

$$RT/[\text{cm}^3\,\text{atm mol}^{-1}]$$
$$= RT/[(\text{J mol}^{-1}) \quad (\text{m}^3\,\text{Pa J}^{-1}) \quad (\text{cm}^3\,\text{m}^{-3}) \quad (\text{atm Pa}^{-1})]$$
$$= 8.3145(T/\text{K}) \times 1 \times 10^{-6} \times 101\,325$$
$$= 82.058(T/\text{K})$$

$$\therefore B/\text{cm}^3\,\text{mol}^{-1} = 34.0 - 3.1 \times 10^6/82.06 \times 333 = \mathbf{-79.4}$$

A6.2 Since $\partial\mu/\partial P = V'$ (equation 5.19):

$$\Delta\mu = \int_1^{80} V'\,\text{d}p = \int_1^{80} (A/P + B)\,\text{d}P$$
$$= A \ln P/\text{atm} - B \times \Delta P$$

$$\therefore \Delta\mu/\text{cm}^3\,\text{atm mol}^{-1} = 22\,410 \ln 80 - 55.6 \times 79 = 93\,800$$

but $1\ \text{cm}^3\,\text{atm} = 10^{-6} \times 101\,325\ \text{m}^3\,\text{Pa} = 0.101\,325\ \text{J}$

$$\therefore \Delta\mu = 0.101\,325 \times 93\,800 = \mathbf{9505\ J\ mol^{-1}}$$

A6.3 Using equation 6.8 with $Z = PV/RT$ and plotting Z *vs.* $\ln P/\text{atm}$, the area between the curve and $Z = 1$ gives the required value for $\ln\gamma$. The area between 0 and 1 atm could be neglected, or estimated from an algebraic expression $PV/RT = 1 - 6 \times 10^{-3} P/\text{atm}$. Indeed, this holds up to 10 atm and gives a value for the area(10 atm) = −0.06 ($\gamma = 0.94$).

Integrals under plotted curves may be obtained by planimeter, by Simpson's rule, by cutting out and weighing, or by counting squares (if the plot is on graph paper). The accuracy is not very high. Approximate results by counting squares:

area(10 atm) = −0.05 or γ = **0.95**, $\therefore f$ = **9.5 atm**

area(50 atm) = −0.36 or γ = **0.70**, $\therefore f$ = **35 atm**

A6.4

$$\int_0^{100} (Z - 1)\,\text{d} \ln P' = \int_0^{100} [aP' + b(P')^2]\,\text{d}P'/P'$$
$$= [aP' + (b/2)(P')^2]_0^{100}$$

with $P' = P/P^\ominus$, $a = -2.38 \times 10^{-3}$, and $b = 5.22 \times 10^{-6}$. The integral is

$$\ln\gamma = -2.38 \times 10^{-1} + 2.61 \times 10^{-2} = 0.212$$

or $\gamma = 0.81$, $\therefore f$ = **81 atm**

A6.5 15 MPa = 150 bar = 150 $P^{\ominus}$.

(a) $$\ln \gamma = 4.1 \times 10^{-4} \times 150 = 0.0615$$

$$\therefore \gamma = \mathbf{1.063},\ f = \mathbf{160\ bar}$$

(b) Assuming the Lewis and Randall rule, $\gamma = f/p = f^*/P$, where f^* is taken at the total pressure, P, of the system (see equation 6.11), the results are the same as in (a).

A6.6 $\Delta_f G = 19.08\ \text{kJ mol}^{-1}$

$$\therefore K^{\ominus} = \exp[-19\,080/(8.314 \times 500)] = 0.010\,15$$

However, to find the per cent conversion, we require K_x.

Let α be the fraction of CO converted into methanol; then at equilibrium, the amounts will be:

	CO	+	$2H_2$	=	CH_3OH	
n/mol	$1 - \alpha$		$2 - 2\alpha$		α	total $N = 3 - 2\alpha$
x	$(1 - \alpha)/N$		$2(1 - \alpha)/N$		α/N	

But

$$K_x = \Pi\, x^{\nu} = (\alpha/N)/[2(1 - \alpha)/N]^2[(1 - \alpha)/N]$$
$$= \alpha(3 - 2\alpha)^2/4(1 - \alpha)^3$$

Now from equation 6.12 we have

$$K_x = (K^{\ominus}/K_{\gamma})(P^{\ominus}/P)^{\Delta\nu}$$

with, in this case, $\Delta\nu = -2$ and $K_{\gamma} = \gamma(CH_3OH,\ 200\ \text{atm})$, since CO and H_2 are 'ideal'.

$$\therefore \ln K_{\gamma} = \ln \gamma(CH_3OH) = \int_0^P (Z - 1)\, d\ln P$$
$$= \int_0^{200} -4 \times 10^{-3}\, dP = -0.8$$
$$\therefore K_{\gamma} = \gamma(CH_3OH) = 0.45$$
$$\therefore K_x = (0.010\,15/0.45)(1/200)^{-2}$$
$$= 902 = \alpha(3 - 2\alpha)^2/4(1 - \alpha)^3$$

To solve the equation for α, a trial and error approach is fairly rapid. $\alpha = 0.5$ gives 4; 0.9, 324; 0.95, 2299; 0.93, 881, which is near enough.

$$\therefore \alpha = \mathbf{93\%}$$

A6.7 (a) From equation 6.14

$$\partial H/\partial P = b - 2a/RT$$
$$= 36 - 2 \times 4.0 \times 10^6/(82.06 \times 750)$$
$$= 36 - 130 = -94 \text{ cm}^3 \text{ mol}^{-1}$$
$$= -94 \times 0.101\,325 = -9.5 \text{ J mol}^{-1}$$

The change in enthalpy is therefore

$$-9.5 \times 39 = \mathbf{-370\ J\ mol^{-1}}$$

From equation 6.15

$$\partial C_P/\partial P = 2a/RT^2 = 130/750 = 0.173 \text{ cm}^3 \text{ mol}^{-1} \text{ K}^{-1}$$

The change in C_P is therefore

$$0.173 \times 0.101\,325 \times 39 = \mathbf{0.7\ J\ K^{-1}\ mol^{-1}}$$

(b) From equation 6.18

$$\mu_{JT} = (2a/RT - b)/C_P = 9.5/(48.1 + 0.7)$$
$$= \mathbf{0.195\ K\ atm^{-1}}$$

(c) $T_{inv} = 2a/Rb = 2 \times 4.0 \times 10^6/(36 \times 82.06) = \mathbf{2708\ K}$

A6.8 For a pure gas $V = V'$ and therefore $V'/RT - 1/P$ becomes $(b - a/RT)/RT$, so that equation 6.7 becomes

$$\ln \gamma = \int_0^P [(b - a/RT)/RT]\, dP$$

This expression vanishes at the lower limit. At the upper limit it is

$$\ln \gamma = [36 - 4.0 \times 10^6/(82.06 \times 750)]P/(82.06 \times 750)P^\ominus$$
$$= 0.954 \text{ for P} = 100 \text{ atm}$$

so that $f =$ **95.4 atm**
(For the given data, $P^\ominus = 1$ atm.)

A6.9 (a) $K_P = K_x P^{\Delta v} = 0.536/(0.116^{0.5} \times 0.348^{1.5} \times 600)$

$$= \mathbf{0.0127}$$

From $\Delta G^\ominus = -RT \ln K^\ominus$

$$\ln K^\ominus = -29\,830/8.314 \times 723 = -4.963$$
$$\therefore K^\ominus = 0.006\,995 = K_f = K_P K_\gamma$$

from which $K_\gamma = 0.551$.

(b) By equation 6.7

$$\ln \gamma = \int_0^P (B + CP + DP^2)\,\mathrm{d}p/RT$$

which gives the required equation when a, b, and c are substituted for B/qRT, $C/2qRT$, and $D/3qRT$, respectively, after integration, and $q = \ln 10$.

(c) The Lewis and Randall rule (equation 6.11) means that γ may be taken as the fugacity of the pure gas at the total pressure of the mixture divided by that pressure, simplifying the calculation considerably. Thus for $P = 600$ atm, substitution in the expression for $\log \gamma$ gives -0.0575 for NH_3, 0.1347 for N_2, and 0.0900 for H_2. But

$$\log K_\gamma = \log \gamma(NH_3) - 0.5 \log \gamma(N_2) - 1.5 \log \gamma(H_2)$$
$$= -0.2598$$

$\therefore$ $K_\gamma =$ **0.550** in close agreement with the value obtained in (a).

Chapter 7

Non-ideal Mixtures

7.1 ACTIVITIES

Unlike gases, liquids rarely form ideal mixtures because of the close proximity of their molecules. Their chemical potentials can, however, be equated to those of their vapours with which they are in equilibrium. Thus, comparing a pure liquid of fugacity f^* with a liquid in a mixture, we get (equation 6.5)

$$\mu^* = \mu^{\ominus}(T) + RT\ln(f^*/P^{\ominus}) \text{ for a pure liquid}$$

and

$$\mu = \mu^{\ominus}(T) + RT\ln(f/P^{\ominus}) \text{ in a mixture}$$

by substitution

$$\mu = \mu^* + RT\ln(f/f^*) \tag{7.1}$$

f/f^* would only equal the mole fraction in the ideal case, but we can preserve the form of the ideal solution equation, equation 5.24, by defining an 'effective' concentration, a, such that

$$a = f/f^* \approx p/p^* \tag{7.2}$$

so that

$$\mu = \mu^* + RT\ln a \tag{7.3}$$

with μ, μ^*, and the *activity*, a, all functions of T, P. Whilst fugacities have been used in the definition of activity, in most cases vapour pressures may be substituted.

As in the case of non-ideal gases, ideal and non-ideal behaviour are related by defining a correction factor called the *activity coefficient:*

$$\gamma = a/x = f/xf^* \tag{7.4}$$

Taking the pure state ($x = 1$) as reference state, then by equation 7.3 $a = 1$ to make $\mu = \mu^*$. Therefore $\gamma = 1$ also. When $a > x$ ($\gamma > 1$), solutions are described as showing positive deviation from Raoult's law, indicating 'repulsion', leading to separation into two immiscible layers in extreme cases, while negative deviation ($\gamma < 1$) indicates mutual attraction between the two components of the mixture, and a tendency to compound formation (with exothermic mixing). Since the 'minority' component is always most sensitive to these interactions, the deviation of γ is always largest at the lowest concentrations, as shown in Figure 7.1.

7.2 HENRY'S LAW

It is possible and convenient sometimes to choose a state other than the pure state as the reference state, as in the case of a solution of a gas or solid in a liquid, where the range of concentrations of the solute is limited, and the pure liquid solute often does not exist. It is usual in this case to take as 'ideal' the linear behaviour in dilute solutions (the tangent at $x = 0$), referred to as Henry's law. Here the reference state would be denoted by $^\ominus$ instead of *. It would be the hypothetical state obtained by extrapolating the 'ideal' line to $x = 1$. At that point $\gamma_H = 1$† and $a_H = f/f^\ominus = \gamma x = 1$, defining the reference state $\mu^\ominus$ in

$$\mu = \mu^\ominus + RT\ln(f/f^\ominus) = \mu^\ominus + RT\ln a_H \tag{7.5}$$

7.3 DETERMINATION OF ACTIVITIES

For volatile substances, activities can be determined from equation 7.2, usually ignoring the deviation of vapours from ideal behaviour.

For non-volatile substances the fugacity must be regarded as a

†The 1987 IUPAC recommendations make a distinction between solute activity coefficients, γ, based on Henry's law, and solvent activity coefficients, 'f', based on Raoult's law, but f is not so used in this book since it also stands for the fugacity. Any necessary distinctions between Raoult's law and Henry's law activities and activity coefficients will be indicated by subscripts R and H to a or γ.

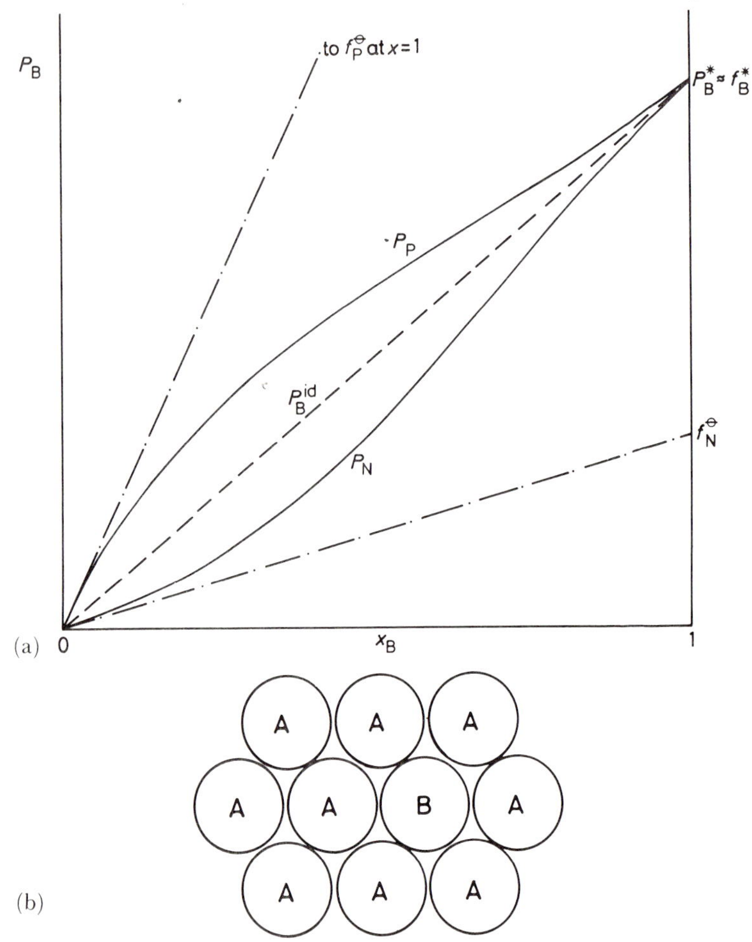

Figure 7.1 *Deviations from Raoult's law (---) and Henry's law (-.-.-.). Figure 7.1a shows the vapour pressure curves, p_P for positive and p_N for negative deviation from Raoult's law (p^{id}). The tangents at zero concentration (infinite dilution) have intercepts $f_P^\ominus$ and $f_N^\ominus$, respectively, at $x = 1$. These are also known as Henry's law constants. Figure 7.1b shows from a simple model why the deviation at infinite dilution is a maximum. It is determined by the number of A–B interactions, which will be a maximum for B surrounded by As, and leads to the hypothetical Henry's law reference state, $f^\ominus$, of pure B held exclusively by A–B interactions. At the other end, A and B will be interchanged, and the escaping tendency of the majority component, B, will be principally influenced by interactions of like molecules. The occasional A–B interaction will only have a marginal effect on the average force field of B, so that the vapour pressure curve will approach ideal behaviour (Raoult's law) tangentially*

purely theoretical concept and other equilibrium phenomena have to be utilized. Thus the activity of an alloy can be determined from the e.m.f. of the cell in which the alloy and pure metal are electrodes in a solution of its salt, and for which

$$\Delta G = \mu - \mu^* = RT \ln a = -n\mathcal{F}\mathcal{E} \tag{7.6}$$

Since the activities replace mole fractions in the chemical potential equation, a must replace x in all expressions dealing with chemical or phase equilibria, such as equilibrium constants, or freezing point depression formulae. For instance, equation 4.9 becomes

$$\ln a_{\mathrm{A}} = (\Delta_{\mathrm{fus}} H_{\mathrm{A}}/R)(T_0^{-1} - T_1^{-1}) \tag{7.7}$$

and allows the activity of a solvent to be determined by freezing point depression, or the activity of a solute by solubility measurements. Both these methods are applicable only to the activity of one component in the binary mixture. However, deviations are based on mutual interactions, and must be in some way related. This is indeed the case, and activities of the other component in a binary mixture can be found from the Gibbs–Duhem equation (equation 5.18). At constant T and P this is

$$n_{\mathrm{A}}\,\mathrm{d}\mu_{\mathrm{A}} = -n_{\mathrm{B}}\,\mathrm{d}\mu_{\mathrm{B}} \tag{7.8}$$

but substituting $\mathrm{d}\mu = RT\,\mathrm{d}\ln a$ from equations 7.3 or 7.5 gives

$$\ln(a_{\mathrm{B},2}/a_{\mathrm{B},1}) = -\int_1^2 (n_{\mathrm{A}}/n_{\mathrm{B}})\,\mathrm{d}\ln a_{\mathrm{A}} \tag{7.9}$$

If the lower limit is the reference state of a_{B}, $a_{\mathrm{B},1} = 1$ and can be omitted. The right-hand side of equation 7.9 can also be rewritten with $(x_{\mathrm{A}}/x_{\mathrm{B}})$ instead of the n ratio, and can then be simplified as follows.

With $\mathrm{d}\ln a = \mathrm{d}\ln\gamma + \mathrm{d}\ln x$, the integral can be split into two with

$$-(x_{\mathrm{A}}/x_{\mathrm{B}})\,\mathrm{d}x_{\mathrm{A}}/x_{\mathrm{A}} = -\mathrm{d}x_{\mathrm{A}}/x_{\mathrm{B}} = \mathrm{d}\ln x_{\mathrm{B}}$$

so that $\ln x_{\mathrm{B}}$ cancels out on both sides of the expression, leaving

$$\ln(\gamma_{\mathrm{B},2}/\gamma_{\mathrm{B},1}) = -\int_1^2 (x_{\mathrm{A}}/x_{\mathrm{B}})\,\mathrm{d}\ln\gamma_{\mathrm{A}} \tag{7.10}$$

or

$$\ln(\gamma_{\mathrm{B}}/\gamma_{\mathrm{B}}^*) = -\int_1^2 (x_{\mathrm{A}}/x_{\mathrm{B}})\,\mathrm{d}\ln\gamma_{\mathrm{A}} \tag{7.11}$$

with the reference state of B as the lower limit.

Equations 7.8—11 are all known as Gibbs–Duhem equations. They can be evaluated graphically or numerically. There are also analytical expressions for $\ln \gamma$, which must themselves be compatible with the Gibbs–Duhem equation. Two of these are the Margules equation:

$$\ln \gamma_1 = [A + 2(B - A)x_1]x_2^2$$

and the van Laar equation:

$$\ln \gamma_1 = AB^2 x_2^2/(Ax_1 + Bx_2)^2 \qquad (7.12)$$

The expression for $\ln \gamma_2$ is obtained by interchanging the subscripts 1 and 2 and the constants A and B.

Both of these 'two-constant' equations converge to the same expression if $A = B$. If, in addition, this sole constant, B, is independent of temperature (unlike A and B in equation 7.12), the new equation is the Hildebrand *regular solution*[6] equation:

$$RT \ln \gamma_1 = Bx_2^2 \qquad (7.13)$$

7.4 DEPENDENCE OF ACTIVITY ON TEMPERATURE AND PRESSURE

In the case of ideal solutions (equation 5.24), differentiation w.r.t. P gave $V' - V^* = 0$. For non-ideal solutions (equation 7.3), we obtain

$$V' - V^* = RT\partial \ln a/\partial P = RT\partial \ln \gamma/\partial P \qquad (7.14)$$

since $\partial \ln \gamma x/\partial P = \partial \ln \gamma/\partial P \quad [\partial \ln x/\partial P = 0]$.

Similarly, division by RT followed by differentiation w.r.t. T gives

$$-(H' - H^*)/RT^2 = \partial \ln a/\partial T = \partial \ln \gamma/\partial T \qquad (7.15)$$

7.5 EXCESS QUANTITIES

Non-ideal mixtures are often described in terms of *excess quantities*, X^E. These are defined as the difference between the relevant quantities for mixing in non-ideal and ideal mixtures. For instance:

$$G^{\mathrm{E}} = \Delta_{\mathrm{mix}} G - \Delta_{\mathrm{mix}} G^{\mathrm{id}} = \Sigma\, xRT(\ln a - \ln x) = \Sigma\, xRT \ln \gamma \tag{7.16}$$

Clearly, G^{E} positive means $\gamma > 1$, or positive deviations from Raoult's law; conversely, $\gamma < 1$ ($\ln \gamma < 0$) leads to $G^{\mathrm{E}} < 0$.

Equations 2.2, 2.7, and 2.8 hold similarly for X^{E}, so that, for instance

$$G^{\mathrm{E}} = H^{\mathrm{E}} - TS^{\mathrm{E}} \text{ or } \partial G^{\mathrm{E}}/\partial T = -S^{\mathrm{E}} \text{ and } \partial G^{\mathrm{E}}/\partial P = V^{\mathrm{E}} \tag{7.17}$$

Applied to regular solutions (equation 7.13):

$$G^{\mathrm{E}} = x_1 B x_2^2 + x_2 B x_1^2 = B x_1 x_2 \quad (\text{since } x_1 + x_2 = 1) \tag{7.18}$$

Since B is independent of T

$$\partial G^{\mathrm{E}}/\partial T = 0,$$

$$\therefore\ S^{\mathrm{E}} = 0\ (viz.\ \Delta S^{\mathrm{reg}} = \Delta S^{\mathrm{id}}) \tag{7.19}$$

Also

$$H^{\mathrm{E}} = G^{\mathrm{E}} = B x_1 x_2 = \Delta_{\mathrm{mix}} H^{\mathrm{reg}} \tag{7.20}$$

[$H^{\mathrm{E}} = \Delta_{\mathrm{mix}} H$ for all non-ideal solutions, since $\Delta_{\mathrm{mix}} H^{\mathrm{id}} = 0$ (equation 5.25)]

EXAMPLES

(In these examples, the pressure, P, stands for the total vapour pressure.)

7.1 The vapour pressures at 50 °C of pure ethanol and water are 222 and 95 torr, respectively. At this temperature and a total pressure of 133 torr, the analysis of the ethanol–water phases in equilibrium showed 4.6 mol % ethanol in the liquid and 29 mol % in the vapour. Compare these figures with the predictions of Raoult's law in terms of the appropriate activities and activity coefficients for the ethanol and water, in the mixture.

7.2 From the data below for isopropanol (1) – benzene (2) mixtures at 298 K, calculate the activity and activity coefficient of isopropanol and benzene at $x_2 = 0.6$

(a) on a Raoult's law basis and
(b) on a Henry's law basis.

x_2	0	0.025	0.05	0.4	0.6	0.9	0.95	1.00
p_2/kPa	0	0.73	1.46	3.80	4.33	5.40	5.57	5.86
p_1/kPa	12.58	12.27	12.06	10.66	9.40	3.93	1.97	0

7.3 The following data refer to the mole fractions in the liquid (x) and vapour (y) phases of water in H_2O–H_2O_2 mixtures. Calculate the activities, with reference to the pure substances, for water and hydrogen peroxide at 75 °C at the given concentrations.

x	0	0.255	0.500	0.800	1.000
y	0	0.601	0.876	0.982	1.000
P/kPa	5.21	9.33	16.68	29.90	38.54

7.4 (a) Show that if a solute obeys Henry's law, the solvent must obey Raoult's law over the same concentration range.
(b) From the equilibrium water vapour pressure data for sucrose solutions at 298 K given below (the last one saturated), find the activity coefficients of sucrose for each of the mole fractions of water tabulated, stating your choice of reference basis for a. (Use equation 7.10 and integrate by counting squares or otherwise.)

10^4x	10^4	9982	9928	9823	9652	9487	9328	9174	8998(sat)
p/Pa	3167	3161	3144	3107	3035	2955	2869	2779	2677

7.5. Derive an equation for $\ln \gamma_2$ for a mixture in which $\ln \gamma_1$ is given by equation 7.13.
If B is independent of T, show that $G^E = H^E = Bx_1x_2$.

7.6 (a) Test the data, given below, for $CCl_4(1) - C_6H_{12}(2)$ mixtures at 40 °C for compatibility with the regular solution equation

$$RT \ln \gamma_2 = Bx_1^2$$

(b) What further data would be required to establish compliance?
(c) Find H^E for the equimolar mixture, assuming these solutions to be regular.

P/mmHg†	184.6	203.4	207.0	210.2	212.1	213.4
x_1	0	0.4739	0.6061	0.7542	0.8756	1.000
y_1	0	0.5103	0.6341	0.7702	0.8822	1.000

7.7 For the Ga–P system containing up to 50 mol % P

$$G^E = \alpha(T)x_P x_{Ga}$$

with $\alpha/\mathrm{J\,mol^{-1}} = -7.53T/\mathrm{K} - 2500$
(a) What are the expressions for H^E and S^E for this system?
(b) How would such a system differ from a 'regular' solution?
(c) Would you expect positive or negative deviations from Raoult's law?

7.8 From the following data for $CHCl_3$ (1) and $(CH_3)_2CO$ (2), calculate the activities and activity coefficients at 308.3 and 323.1 K, stating the reference basis. Calculate $\Delta_{mix}G$ at both temperatures and estimate $H' - H^*$ for both components at $x = 0.5$; hence estimate $\Delta_{mix}H$.

T/K	p_1^*/torr	p_2^*/torr	P_{total}/torr	x_1(liq)	y_1(vap)
308.3	303	332	255	0.500	0.439
323.1	521	612	469	0.500	0.450

7.9 From the results obtained in example 7.8 find the entropy of mixing in that system and compare it with the ideal value.

ANSWERS

[In the whole of this section the following assumptions may be made: (a) equation 7.2, $a_R \approx p/p^*$ to a sufficient approximation; (b) the partial pressure of any vapour component, B, is $p_B = y_B P$, where y is the mole fraction of the vapour and P the total pressure.]

A7.1 For ethanol

$$p_e = y_e P,\ a_e = p_e/p_e^*,\ \text{and}\ \gamma_e = a_e/x_e$$

$$a_e = 0.29 \times 133/222 = \mathbf{0.174}, \quad \gamma_e = 0.174/0.046 = \mathbf{3.78}$$

$$a_w = 0.71 \times 133/95 = \mathbf{0.994}, \quad \gamma_w = 0.994/0.954 = \mathbf{1.04}$$

†P is the total vapour pressure of the mixture.

A7.2 (a) On the basis of Raoult's law

$$a_R = p/p^* \text{ and } \gamma_R = a_R/x$$

For benzene: $a_R = 4.33/5.86 = \mathbf{0.739}$,

$$\gamma_R = 0.739/0.6 = \mathbf{1.23}$$

For isopropanol: $a_R = 9.40/12.58 = \mathbf{0.747}$,

$$\gamma_R = 0.747/0.4 = \mathbf{1.87}$$

(b) On the basis of Henry's law, $a_H = p/f_H^\ominus$, where $f_H^\ominus$ is the value of f (or in this case, p) at the intersection of the tangent at $x = 0$ with the line $x = 1$. The slope of the tangents can be found by inspection from the initial linear portion of the plots.
Extrapolation to $x = 1$ from 0 . . 1.97 . . 3.93 (≈3.94) gives $f_H^\ominus = 39.4$ kPa for isopropanol

$$\therefore a_H = 9.40/39.4 = \mathbf{0.239}$$

$$\gamma_H = 0.239/0.4 = \mathbf{0.598}$$

Similarly for benzene $f_H^\ominus = 29.2$ kPa

$$\therefore a_H = 4.33/29.2 = \mathbf{0.148}$$

$$\gamma_H = 0.148/0.6 = \mathbf{0.247}$$

7.3 Again, $p = yP$, so that

$$a_R = p/p^* = yP/p^*$$

with $p^* = 38.54$ kPa for water and 5.21 kPa for hydrogen peroxide.

Inserting these values for	$x =$	0.255	0.500	0.800
gives for water	$a_R =$	**0.146**	**0.379**	**0.762**
and for H_2O_2	$a_R =$	**0.715**	**0.397**	**0.104**

A7.4 The activity coefficient γ_1 for water is readily calculated from $\gamma = p/xp^*$ for water, and x_1/x_2 can be plotted *vs.* $\ln \gamma$ (see Figure 7.2). The areas of the segments under the curve are then the values of $\Delta \ln \gamma_2$ according to the Gibbs–Duhem equation. They must be used (with their appropriate signs) to calculate successive $\ln \gamma_2$ values.

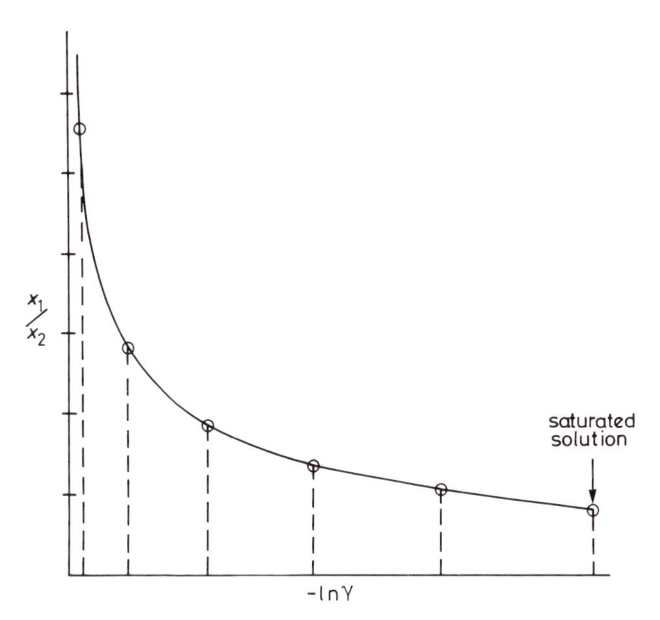

Figure 7.2 *Determination of the activity coefficient of a solute in solution from the activity of the solvent, using the Gibbs–Duhem equation, equation 7.11. Areas are taken from the reference state of 2 to each desired value of* x_2. *The same method could also be used for a instead of* γ *(Adapted from ref. 7)*

The simplest reference state to adopt is pure solid sucrose, which also makes the activity of sucrose in the saturated solution 1, but its mole fraction is $1 - 0.8998 = 0.1002$, so that $\gamma_2 = a/0.1002 = 9.98$. The lower limit in the integral of equation 7.10 is the saturated solution. Thus, starting with $\ln 9.98 = 2.300$ and $\Delta \ln \gamma_2 = 0.182$, the next value of $\ln \gamma_2$ is $2.300 - 0.182 = 2.118$, and $\gamma_2 = 8.32$.

For the figures calculated below, each area shown is that for the trapezium and therefore likely to be too large, but less liable to error than that obtained from counting squares. Since for the last segment $x_2 = 0$ makes x_1/x_2 infinite, its area can only be estimated by, for instance, assuming the regular solution equation for that section and using an algebraic integration. This could also be done if Henry's law activity coefficients (based on infinite dilution) are required. The two sets, γ_R and γ_H, are related by a constant factor.

$10^4 x$	10^4	9982	9928	9823	9652	9487	9328	9174	8998 (sat)
p/Pa	3167	3161	3144	3107	3035	2955	2869	2779	2677
$-10^4 \ln \gamma_1$	0.00	0.00	0.63	13.7	71.6	165	292	444	625
x_1/x_2	∞	555	138	55.5	27.7	18.5	13.9	11.1	8.98
$10^3 \Delta$ area		(3.3)	20.8	126	241	217	205	190	182
γ_2	**(3.05)**	**3.06**	**3.13**	**3.55**	**4.51**	**5.61**	**6.88**	**8.32**	**9.98**

A7.5 Since $RT \ln \gamma_1 = Bx_2^2$

$$\mathrm{d} \ln \gamma_1 = (B/RT) 2x_2 \, \mathrm{d}x_2$$

which substituted in equation 7.11 gives

$$\ln \gamma_2 = -\int (x_1/x_2)(B/RT) 2x_2 \, \mathrm{d}x_2$$

$$= -\int (B/RT) 2x_1 (-\mathrm{d}x_1) = \boldsymbol{(B/RT)x_1^2 + I}$$

Since γ_1 in equation 7.13 is the Raoult's law activity coefficient, $\gamma_1 = 1$ for pure 1, *i.e.* at $x_2 = 0$. Similarly, for Raoult's law $\gamma_2 = 1$ for pure 2, so that the integration constant, I must be 0. For Henry's law, *i.e.* infinite dilution as reference state, I would have to become $-B/RT$ to satisfy the boundary condition.

See equation 8.3 for the factor connecting a_R and a_H, and hence γ_R and γ_H.

A7.6 (a) $\gamma_2 = p_2/p_2^* = (1 - y_1)P/[184.6(1 - x_1)]$

$\ln \gamma_2 / x_1^2 = B/RT$ must be constant within the limits of accuracy. The calculated values are shown in the table below. The values for γ at $x = 0$ are indeterminate (0/0) and could only be found by some extrapolation. The last line of the table was calculated using $B/RT = 0.1100$, but the average value would also give deviations of less than 1 in 10^3 for the 'regular' value of γ from that in line 2 and may be regarded as satisfactory agreement.

x_1	0	0.4739	0.6061	0.7542	0.8756	1.000
γ_2	1	1.0256	1.0416	1.0645	1.0880	–

$\ln\gamma_2/x_1^2$	–	0.1125	0.1110	0.1099	0.1100
$(\gamma_2)_{reg}$	–	1.0252	1.0412	1.0646	1.0880

(b) B would have to be independent of T. To establish this, data at a different temperature would be required.

(c) The value 0.1100 for B/RT with $H^E = Bx_1x_2$ gives **71.6 J mol⁻¹**.

A7.7 (a) $G^E = \alpha(T)x_P x_{Ga}$, with $\alpha/\text{J mol}^{-1} = -7.53T/\text{K} - 2500$

$$\therefore S^E = -\partial G^E/\partial T = \boldsymbol{-x_P x_{Ga}\,\partial\alpha/\partial T},$$

with

$$\partial\alpha/\partial T = -7.53\ \text{J K}^{-1}\ \text{mol}^{-1}$$

$$\therefore H^E = G^E + TS^E = \mathbf{-2500\ J\ mol^{-1}}$$

(b) B for regular solutions is independent of T, whereas $\alpha(T)$ is not. $\Delta H = H^E$ is negative, implying attractive forces between P and Ga. This would lead to clustering, which is incompatible with the 'ideal' entropy of regular solution. For that reason, regular solutions must have positive values of H^E (see section 7.5).

(c) The given α makes G^E negative for all x, but $G^E = \Sigma xRT\ln\gamma$, which can only become negative if $\ln\gamma$ is; therefore $\gamma < 1$, which means deviations from Raoult's law are negative (see equation 7.16).

A7.8 $\gamma_1 = y_1P/(x_1p_1^*)$ and $\gamma_2 = (1 - y_1)P/[(1 - x_1)p_2^*]$

At 308.3 K

$\gamma_1 = 0.439 \times 255/0.5 \times 303 =$ **0.739** and $\ln\gamma_1 = -0.302$

$\gamma_2 = 0.561 \times 255/0.5 \times 332 =$ **0.862** and $\ln\gamma_2 = -0.149$

At 323.1 K

$\gamma_1 = 0.450 \times 469/0.5 \times 521 =$ **0.810** and $\ln\gamma_1 = -0.211$

$\gamma_2 = 0.550 \times 469/0.5 \times 612 =$ **0.843** and $\ln\gamma_2 = -0.171$

$$\Delta_{mix}G = \Sigma xRT\ln a$$

$$= 0.5 \times 8.314 \times 323.1 \times (-0.904 - 0.864)$$

$$= \mathbf{-2374\ J\ mol^{-1}}$$

At 308.3 K

$$\Delta G = 0.5 \times 8.314 \times 308.3 \times (-1.837) = -2355 \text{ J mol}^{-1}$$

By equation 7.15

$$\partial \ln \gamma / \partial T = -(H' - H^*)/RT^2$$

$$\therefore H_1' - H_1^* = -RT^2 \, \partial \ln \gamma / \partial T$$

$$= -RT^2 \times (0.091/14.8)$$

$$= \mathbf{-5090 \text{ J mol}^{-1}}$$

taking a mean temperature ($T^2 = 308.3 \times 323.1$).

$$H_2' - H_2^* = -8.314 \times 308.3 \times 323.1 \times (-0.022/14.8)$$

$$= \mathbf{1230 \text{ J mol}^{-1}}$$

[*N.B.* $\delta \ln \gamma$ and δT must be taken in the same direction.]

$$\Delta_{\text{mix}} H = \Sigma x (H' - H^*)$$

$$= 0.5 \times (-5090 + 1230)$$

$$= \mathbf{-1930 \text{ J mol}^{-1}}$$

A7.9 By equation 2.8

$$\Delta S = -\partial \Delta G / \partial T$$

so that

$$\Delta_{\text{mix}} S \approx -(-2374 + 2355)/(323.1 - 308.3)$$

$$= \mathbf{1.28 \text{ J K}^{-1} \text{ mol}^{-1}}$$

The ideal entropy of mixing would be

$$\Delta_{\text{mix}} S^{\text{id}} = -\Sigma x \ln x = -\ln 0.5 = 5.76 \text{ J K}^{-1} \text{ mol}^{-1}$$

The low actual value for $\Delta_{\text{mix}} S$ (S^{E} negative) again points to some ordering in the mixture due to the attractive forces evidenced by the negative value of ΔH (*cf.* remark in answer A7.7b).

Chapter 8

General Treatment of Equilibrium

8.1 EQUILIBRIUM AND CHEMICAL POTENTIALS

Having studied chemical potentials in ideal and non-ideal systems for mixtures as well as pure compounds, we can now show the simplicity of the chemical potential approach to both chemical and phase equilibrium studies; for either, $\Delta G = 0$ at constant T and P. The main difference is that physical changes involve infinitely variable ratios of the amounts of reactants, whereas in chemical changes they are normally fixed. We therefore express ΔG as $\Sigma\, n\mu$ for one and $\Sigma\, \nu\mu$ for the other.

All that remains is to put in the appropriate form of μ, taking account of the reference data available and what approximations can be tolerated. Thus, in the general case of a chemical reaction

$$\Delta G = \Sigma\, \nu(\mu^{\ominus} + RT \ln k)$$

where k is a numeric variable such as $p/P^{\ominus}$, $f/P^{\ominus}$, x, or a, which relates the actual state to the reference state, $\mu^{\ominus}$, and $\Sigma\, \nu\mu^{\ominus} = \Delta G^{\ominus}$, the standard Gibbs energy. The remaining terms $\Sigma\, RT \ln k^{\nu}$ form the correction term, $RT \ln Q$, so that

$$\Delta G = \Delta G^{\ominus} + RT \ln Q \tag{8.1}$$

At equilibrium, $\Delta G = 0$ and the general term Q becomes the equilibrium constant $K^{\ominus}$, given by

$$RT \ln Q_{\text{eq}} = RT \ln K^{\ominus} = -\Delta G^{\ominus} \tag{8.2}$$

As an example, take a reaction in which a gas, a liquid, and solids participate:

$$H_2(g) + 2AgCl(s) = 2Ag(s) + 2HCl(aq)$$

Q or $K^{\ominus}$ for this reaction would be

$$(a_{\mathrm{HCl}})^2/(p/P^{\ominus})_{\mathrm{H_2}}$$

But for $K^{\ominus}$, a_{HCl} and $p_{\mathrm{H_2}}$ would have to have their equilibrium values and for Q, the initial value of $p_{\mathrm{H_2}}$ and the final value of a_{HCl}. The expression would not contain any term for either of the two solids, Ag or AgCl, since they are in their reference states, which would only enter into the $\Delta G^{\ominus} = \Sigma \nu\mu^{\ominus}$ part. Even if these solids are in contact with water in which they have some minute solubility, their chemical potentials would not change as long as there is any of the solid left.

There are several different reference states possible for the HCl in this example:

(a) $a_{\mathrm{R}} = p/p^*$, where p^* would be the v.p. of liquid HCl at the same temperature. This a would be the activity referenced to Raoult's law.
(b) $a_{\mathrm{H}} = p/f^{\ominus}$, where $f^{\ominus}$ is defined in section 7.2 (Henry's law), *viz.* where the tangent at $x = 0$ to the vapour pressure *vs.* mol fraction curve intersects the line $x = 1$ (see Figure 7.1).

$$\therefore f^{\ominus} = (\partial p/\partial x)_{x=0}$$

(c) $a_{\mathrm{m}} = p/f^{\ominus}_m$, where a similar tangent (to p *vs.* m) intersects the line $m = m^{\ominus}$, the line of unit molality.†

$$\therefore f^{\ominus}_m = [\partial p/\partial(m/m^{\ominus})]_{m=0}$$

The three different activities can be readily related through the pressure (here assumed to be equal to the fugacity):

$$p = a_{\mathrm{R}}p^* = a_{\mathrm{H}}f^{\ominus} = a_m f^{\ominus}_m$$

$$\therefore \quad a_{\mathrm{R}} : a_{\mathrm{H}} : a_m = 1/p^* : 1/f^{\ominus} : 1/f^{\ominus}_m \tag{8.3}$$

The values of $f^{\ominus}$ and $f^{\ominus}_m$ are simply related:

$$[\partial p/\partial(m/m^{\ominus})]_{m=0} = (\partial p/\partial x)_{x=0}[\partial x/\partial(m/m^{\ominus})]_{m=0}$$

But $x = m/(M^{-1} + m)$ and for water as solvent, $M^{-1} = 55.51\ \mathrm{mol\,kg^{-1}}$.

$$\therefore [\partial x/\partial(m/m^{\ominus})]_{m=0} = 1/55.51$$

so that for aqueous solutions $f^{\ominus} = 55.51 f^{\ominus}_m$. $f^{\ominus}_m$ is the reference

†$m^{\ominus} = 1\ \mathrm{mol\,kg^{-1}}$; see also section 4.2.

state usually adopted for electrolyte solutions in water, except that allowance has to be made for the splitting into ions. Thus $HCl = H^+ + Cl^-$ has an equilibrium constant $K = a_+a_-/a_{HCl}$. But since individual ion activities cannot be measured, they are replaced by the mean ionic activity, so that $K = (a_\pm)^2/a_{HCl}$.

Such ionic activities feature in chemical cells, such as

$$Ag|AgCl|HCl(m)|H_2|Pt$$

with cell reaction $Ag + HCl = AgCl + \frac{1}{2}H_2$ or in concentration cells, such as two such cells back to back:

$$Ag|AgCl|HCl(m_1)|H_2, Pt|HCl(m_2)|AgCl|Ag$$

in which the driving force is the difference in the HCl concentrations:

$$\Delta G = RT \ln a_2/a_1 = -n\mathcal{F}\mathcal{E}$$

which allows the direct determination of electrolyte activities from a series of measurements down to low concentrations.

8.2 SOLUBILITY

Solubility implies a saturated solution, *i.e.* pure solute in one phase in equilibrium with another phase containing the solute and a solvent. The chemical potentials of the solute must then be equal for two phases in equilibrium. This is much simpler than the method used in section 4.2 and requires no assumptions as to volatility. Thus, for the solubility of a solid in a liquid, we have, at saturation, that

$$\mu_s^* = \mu_{soln} = \mu_l^* + RT \ln k \qquad (8.4)$$

with $k = x_{sat}$ (ideal) or a_{sat} (non-ideal). Division by RT and differentiation (see equation 2.8) gives

$$\partial \ln x_{sat}/\partial T = -(H_s^* - H_l^*)/RT^2 = \Delta_{fus}H/RT^2 \qquad (8.5)$$

which can then be integrated between the melting point of the solid and any other temperature to give equation 4.11, if $\Delta_{fus}H$ is assumed independent of T. (The same procedure holds for $k = a_{sat}$.)

This method may also be used for the ideal solubility of a gas, but it must be realized that in these cases the criteria of 'ideal mixing' as in equation 5.25 can only apply to the mixing of the 'liquefied' solid or gas in the liquid mixture. Similarly, one can consider the ideal solubility of a liquid in a gas (*e.g.* the humidity), where the ideal mixture would be between the vaporized liquid and the gas.

Ideal solubility of a gas in a liquid can also be treated as an application of Raoult's law ($x_{sat} = p/p^*$). In this case, p is the pressure exerted by the gas in question, and p^* is the v.p. of the pure liquefied gas at the same T.

Where solutions are non-ideal, the equality of chemical potentials is more likely to be used in the determination of activity coefficients from measured solubilities.

All these applications are best understood through examples of actual cases. These are discussed below.

EXAMPLES

The data for examples 8.1—8.3, given here, are all for 1873 K.

$$C + \tfrac{1}{2}O_2 = CO, \qquad \Delta G^\ominus = -65.93\ \text{kcal mol}^{-1}$$

$$CO + \tfrac{1}{2}O_2 = CO_2, \qquad \Delta G^\ominus = -28.82\ \text{kcal mol}^{-1}$$

$$Fe(l) + \tfrac{1}{2}O_2 = FeO(l), \qquad \Delta G^\ominus = -34.47\ \text{kcal mol}^{-1}$$

$$H_2 + \tfrac{1}{2}O_2 = H_2O(g), \qquad K^\ominus = 1.000 \times 10^4$$

8.1 From the data given above show that the activity (relative to pure solid graphite) of carbon in solution in liquid iron at 1873 K and in equilibrium with a gaseous mixture of CO_2 and CO can be represented by the equation

$$\log_{10} a_C = \log_{10}[p_{CO}^2/(P^\ominus p_{CO_2})] - 4.330$$

The solubility of graphite in liquid iron at 1873 K is 5.4 wt%, and the $p_{CO}^2/(P^\ominus p_{CO_2})$ ratio is 1000 for a 1.8 wt % solution. Compare a_C obtained for the 1.8 wt% solution with approximate predictions from Henry's law.

8.2 A melt containing 0.21 mol% of FeO dissolved in liquid iron at 1873 K was in equilibrium with H_2(g) and H_2O(g) such that

$$p_{H_2O}/p_{H_2} = 0.260$$

Calculate the activity and activity coefficient of FeO in this melt.

8.3 (a) If FeO(l) in Fe(l) at 1873 K obeys Henry's law right up to saturation at 0.8 mol% of FeO, what would be the concentration of FeO in Fe in equilibrium with a slag of equimolar mixture of FeO and MnO, which may be assumed to form an ideal mixture?
(b) If SiO_2 is substituted for MnO in (a), what effect would be expected on the activity of the FeO in the liquid Fe, taking into account that SiO_2 and FeO mix exothermically?

8.4 An iron melt containing 0.8 mol% FeS at 1800 K was found to be in equilibrium with a mixture of H_2 and H_2S such that

$$p_{H_2S}/p_{H_2} = 6.5 \times 10^{-4}$$

(a) Calculate the activity of FeS in this melt, stating the reference basis.

(b) Assuming Henry's law is obeyed by solutions of FeS in Fe at low concentrations, estimate the activity of FeS in a melt containing 1 wt% S, (= 1.75 mol%), sometimes used as a reference basis, and determine ΔG_{1800} for the reaction (occurring in a large bulk of the melt)

$$\mathrm{Fe(l)} + \tfrac{1}{2}\mathrm{S_2(g)} = \mathrm{FeS(1\ wt\%\ S\ in\ liquid\ Fe)}$$

$$\Delta_f G(\mathrm{FeS})/\mathrm{J\,mol^{-1}} = -178 \times 10^3 + 61T$$

$$\Delta_f G(\mathrm{H_2S})/\mathrm{J\,mol^{-1}} = -91 \times 10^3 + 50T$$

8.5 The vapour pressure of pure water is 23.8 torr at 298 K. The vapour pressure of water in equilibrium with $Na_2CO_3{\cdot}H_2O$ and Na_2CO_3(anhydr) is given by the expression

$$\log_{10} p/\mathrm{torr} = 10.825 - 3000(T/\mathrm{K})^{-1}$$

(a) Calculate the standard Gibbs energy change for the reaction

$$\mathrm{Na_2CO_3 + H_2O(g) = Na_2CO_3{\cdot}H_2O}$$

(b) At 298 K, what would be the activities of water in benzene saturated with water, and after drying with excess anhydrous sodium carbonate ?

8.6 A small quantity of 90 mol% aqueous ethanol was allowed to stand in contact with an excess of $CaSO_4{\cdot}\frac{1}{2}H_2O$. It may be assumed that some $CaSO_4{\cdot}2H_2O$ is formed. When equilibrium was established at 298 K, the concentration of ethanol had increased to 90.6 mol%.

Calculate the activity and activity coefficient of water in the equilibrium mixture at 298 K.

The figures below are $\Delta_f G/\text{kJ mol}^{-1}$ for 298 K.

$CaSO_4 \cdot \frac{1}{2}H_2O$	$CaSO_4 \cdot 2H_2O$	$H_2O(g)$	$H_2O(l)$
−1435	−1796	−228.582	−237.141

8.7 (a) If iron melts at 1809 K with the absorption of 15.36 kJ mol^{-1} and C_p of the liquid exceeds that of the solid by 1.25 J K^{-1} mol^{-1}, show that for Fe(s) = Fe(l)

$$\Delta G/\text{J mol}^{-1} = 13100 - 1.25T' \ln T' + 2.14T'$$

with $T' = T/\text{K}$.

(b) At 1673 K, a liquid mixture of FeS in 87 mol% Fe is in equilibrium with almost pure solid Fe. Estimate the activity of the iron in the melt, stating the reference basis.

8.8 The following e.m.f. measurements were obtained from the cell

$$\text{Zn}|\text{ZnSO}_4(\text{soln})|\text{Zn–Cd}(x_{\text{Zn}} = 0.15)$$

$\mathcal{E}$/mV	27.5	29.4	36.7	39.6
T/K	709	737	814	845

(a) Derive the activities and activity coefficients for Zn in the Zn–Cd alloy at the given temperatures and state the reference basis.

(b) Estimate the heat of reaction when 1 mole of Zn is added to a large amount of 15 mol% Zn–Cd alloy.

8.9 Use a treatment analogous to equations 8.4 and 8.5 to arrive at the equation for the ideal solubility (y for a gas) of a liquid in a gas

$$\ln y_{\text{sat}} = -(\Delta_{\text{vap}}H/R)(T^{-1} - T_0^{-1}), \text{ with } T_0 = T_{\text{b.p.}}$$

[Note that this becomes identical with the results of the Clausius–Clapeyron equation (equation 4.6) with y_{sat} for p/p_0, subscript zero referring to b.p.]

Calculate y_{sat} for water at 313 K with a mean $\Delta_{\text{vap}}H$ = 41.90 kJ mol^{-1}. Compare it with the observed v.p., p = 55.13 torr at 313 K.

8.10 Calculate the 'ideal solubility' of C_2H_2 at 273 K in a liquid, assuming its $\Delta_{vap}H$ to be $16.70\ \mathrm{kJ\,mol^{-1}}$ at its normal boiling point of 189 K.

ANSWERS

A8.1 This is an application of the procedure outlined in section 8.1. For $C(soln) + CO_2 = 2CO$

$$\Delta G^{\ominus} = -65.93 + 28.82 = -37.11\ \mathrm{kcal\,mol^{-1}}$$

Treating CO and CO_2 as ideal gases (equation 5.23) and C(soln in Fe) as a non-ideal solution (equation 7.3):

$$\Delta G = 2\mu^{\ominus}_{CO} - \mu^{\ominus}_{CO_2} - \mu_C$$
$$+ RT[2 \ln k_{CO} - \ln k_{CO_2} - \ln a_C]_{eqm} = 0$$
$$\therefore K^{\ominus} = \exp(-\Delta G^{\ominus}/RT)$$
$$= (p_{CO})^2/P^{\ominus} p_{CO_2} a_C \text{ (see section 8.1)}$$

where $k_{CO} = p_{CO}/P^{\ominus}$, *etc.* and $\Sigma\, \nu\mu = \Delta G^{\ominus}$.

$$\therefore \ln K^{\ominus} = 2.303 \log K^{\ominus} = (37\,110 \times 4.184/8.314 \times 1873)$$
$$= 9.971 = 2.303 \times 4.330$$

which gives the required equation on re-arrangement.

For a 1.8 wt% solution

$$\log a_C = 3.000 - 4.330 = -1.330,\ \therefore a_C = \mathbf{0.047}$$

But at 5.4 wt%, the solution is saturated, *i.e.* in equilibrium with pure C, so that $a_C = 1$, and if Henry's law is obeyed (in this context Henry's law is merely a straight line down to $x = 0$), at 1/3 that concentration **it would be 1/3**. Actually mol% and not wt% should be used, but this does not improve the agreement with the observed activity. Its very low value points to some specific Fe–C interaction which holds the C atoms strongly at low concentrations.

A8.2 The reaction involved is

$$FeO + H_2 = Fe + H_2O,\ \Delta G^{\ominus} = \Sigma\, \Delta_f G$$

and the corresponding K ratio is

$$K^{\ominus} = (p_{H_2O}/p_{H_2})(a_{Fe}/a_{FeO})$$

$$K^{\ominus}_{FeO} = \exp(-\Delta_f G/RT)$$

$$= \exp(34\,470 \times 4.184/8.314 \times 1873)$$

$$= 1.053 \times 10^4$$

$$a_{FeO} = (p_{H_2O}/p_{H_2})a_{Fe}/K^{\ominus}$$

with

$$K^{\ominus} = 1.000 \times 10^4/1.053 \times 10^4$$

a_{Fe} can only be estimated on the basis of $\gamma = 1$, since a_R approaches x tangentially at high concentations.

$$\therefore\ a_{Fe} \approx 1 - x_{FeO} \approx 0.998$$

$$\therefore\ a_{FeO} = 1.053 \times 0.26 = \mathbf{0.274}$$

and

$$\gamma = 0.274/0.0021 = \mathbf{130}$$

[The reference bases for Fe and FeO must be the respective standard states of the $\Delta G^{\ominus}$ used, *viz.* pure Fe(l) and FeO(l).]

A8.3 (a) Again, the starting point is that, at saturation, the activity of FeO must be 1 in both phases in equilibrium, *viz.* pure FeO(l), and FeO in the Fe melt. If, however, the activity in the slag is reduced to 0.5 (equimolar ideal mixture), a in the metal phase must follow and so the concentration of oxide must fall to $0.5 \times 0.8 =$ **0.4 mol%**, if Henry's law is obeyed.

(b) ΔH negative would imply some bonding and might suggest a lowering of the activity. Care is necessary, however, particularly at high temperatures, not to ignore the $-T\Delta S$ term in ΔG. In this instance, if a 1 : 1 compound were to be formed, A + B = AB has a negative ΔS instead of a positive $\Delta_{mix}S$, and this could offset a small ΔH term. ΔG would then become positive and $\gamma > 1$.

A8.4 (a) $FeS + H_2 = Fe + H_2S$, with $\Delta G = \Delta_f G(H_2S) - \Delta_f G(FeS)$

$$\therefore \Delta G = 87 \times 10^3 - 11T = 67.2 \text{ kJ mol}^{-1} \text{ at } 1800 \text{ K}$$

Also

$$\ln K^{\ominus} = -67\,200/8.314 \times 1800 = -4.49 = \ln 0.0112$$

But $K^{\ominus} = (a_{Fe}/a_{FeS})/(p_{H_2S}/p_{H_2})$ with the p ratio given and $a_{Fe} \approx x_{Fe} = 1 - x_{FeS} = 0.992$ (for almost pure Fe)

$$\therefore a_{FeS} = a_{Fe} \times 6.5 \times 10^{-4}/0.0112$$

$$= 0.992 \times 0.0580 = \mathbf{0.0575}$$

(b) If Henry's law applies

$$a_{FeS}(1\%) = 0.0575 \times 1.75/0.8 = 0.126$$

But for FeS(pure) $\rightarrow$ FeS(a)

$$\Delta G = \mu - \mu^{\ominus} = RT \ln a = -31 \text{ kJ mol}^{-1}$$

$\therefore$ for $Fe + 1/2S_2 = FeS(1\%\ S)$,

$$\Delta_f G = -68.2 - 31 = \mathbf{-99.2\ kJ\ mol^{-1}}$$

A8.5 (a)

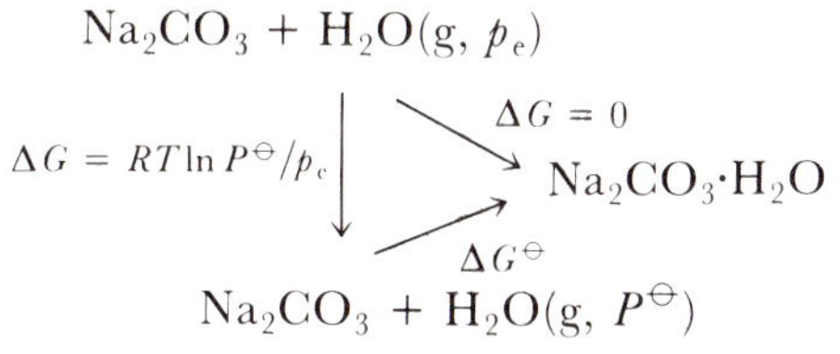

$$Na_2CO_3 + H_2O(g, p_e)$$

$\Delta G = RT \ln P^{\ominus}/p_e$ (downward arrow); $\Delta G = 0$ (arrow to $Na_2CO_3{\cdot}H_2O$); $\Delta G^{\ominus}$ (arrow from lower to $Na_2CO_3{\cdot}H_2O$)

$$Na_2CO_3{\cdot}H_2O$$

$$Na_2CO_3 + H_2O(g, P^{\ominus})$$

When the v.p. is the equilibrium pressure p_e, $\Delta G = 0$
When the v.p. is the standard pressure $P^{\ominus}$, $\Delta G = \Delta G^{\ominus}$
But following the arrows

$$\Delta G = RT \ln P^{\ominus}/p_e + \Delta G^{\ominus} = 0$$

so that

$$\Delta G^{\ominus} = RT \ln p_e/P^{\ominus}$$

[As an alternative approach write down the $\Delta G = \Sigma\, \nu\mu$ equation for this reaction, bearing in mind that the solids only have $\mu^{\ominus}$ terms which only enter into $\Delta G^{\ominus}$, *i.e.*

$$\Delta G = \mu^{\ominus}(\text{hydr}) - \mu^{\ominus}(\text{anhydr}) - \mu^{\ominus}(\text{H}_2\text{O}) - RT\ln p/P^{\ominus}$$

$$= \Delta G^{\ominus} + RT\ln Q$$

At equilibrium, $\Delta G \to 0$, $Q = P^{\ominus}/p$ becomes $K^{\ominus} = P^{\ominus}/p_e$, giving

$$\Delta G^{\ominus} = -RT\ln K^{\ominus} = RT\ln p_e/P^{\ominus}$$

as before (see equations 8.1 and 8.2).]

But $\ln p/P^{\ominus} = \ln p/\text{torr} + \ln \text{torr}/P^{\ominus}$ with $P^{\ominus} \approx 750$ torr.

$$\therefore \Delta G^{\ominus} = RT\ln p/P^{\ominus}$$

$$= \mathbf{2.303\ \mathit{RT}[7.950 - 3000(\mathit{T}/K)^{-1}]}$$

(b) Both μ and a_R are the same for phases in equilibrium and so, neglecting the small concentration of benzene in the water, $\boldsymbol{a_R \approx 1}$.

(c) After drying, the vapour pressure of water would be equal to the equilibrium pressure of the Na_2CO_3–$Na_2CO_3{\cdot}H_2O$ system, 5.73 torr at 298 K [from the equation]. $\therefore a = p/p^* = 5.73/23.8 =$ **0.24**

A8.6 $CaSO_4{\cdot}\frac{1}{2}H_2O + \frac{3}{2}H_2O(g) = CaSO_4{\cdot}2H_2O$
As in A8.5 above:

$$\Delta G^{\ominus} = \tfrac{3}{2}RT\ln p_e/P^{\ominus} = -1796 + 1435 + 1.5 \times 228.6$$

$$= -18\ \text{kJ mol}^{-1}$$

$$\therefore p_e/P^{\ominus} = \exp(-2/3 \times 18\,000/8.314 \times 298)$$

$$= 7.9 \times 10^{-3}$$

$$\therefore p_e = 5.9\ \text{torr}$$

From

$$H_2O(g, P^{\ominus}) \to H_2O(g, p^*) = H_2O(l),$$

$$\Delta G^{\ominus} = -8559\ \text{J mol}^{-1}$$

it follows that

$$\ln p^*/P^{\ominus} = -8559/8.314 \times 298 = -3.455$$

$$\therefore p^* = 23.7\ \text{torr}$$

Since $a = p/p^* = 5.9/23.7 =$ **0.25**, so that $\gamma = 0.25/0.094$ $=$ **2.7**.

A8.7 (a) $\Delta_{fus}H_T = a + b(T - T^{\ominus})$ with $a = \Delta H_{T^{\ominus}}$, $b = \Delta C_P$, and $T^{\ominus} = 1809$ K

But

$$\Delta G/T = -\int \Delta H/T^2\,dT = (a - bT^{\ominus})/T - b\ln T + I$$

or

$$\Delta G = a - bT^{\ominus} - bT\ln T + IT$$

with the integration constant I chosen to make $\Delta G = 0$ at 1809 K

(b) At 1673 K

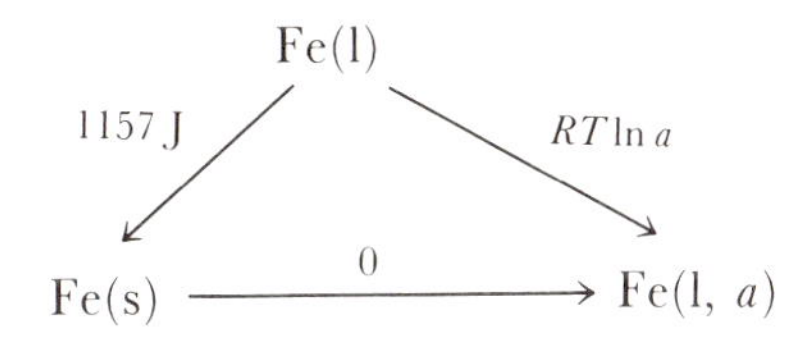

$\therefore$ $RT\ln a = -1157$ J mol^{-1} and $a =$ **0.92** relative to pure supercooled Fe(l, 1673 K). [As an alternative approach, (b) could be regarded as an application of the depression of the freezing point of Fe for the determination of activities (see equation 4.9) or as the 'solubility' of Fe in FeS(l) in terms of equation 8.5, but taking the variation of ΔH with T into account. This will yield exactly the same equation.]

A8.8 (a) The driving force for this electrode concentration cell is the transfer of Zn from the left-hand electrode (via the solution) to the right-hand alloy electrode. It is therefore a process of solution of Zn in Zn-alloy.
$\therefore$ $\Delta G = \mu - \mu^{\ominus} = RT\ln a$, but it is also $-n\mathcal{F}\mathcal{E}$, where $n\mathcal{F}$ is the amount of charge involved ($2\mathcal{F}$ for Zn^{2+}).
$\therefore$ $\ln a = -2\mathcal{F}\mathcal{E}/RT$.

This gives the following results (reference basis pure Zn):

$\mathcal{E}$/mV	27.5	29.4	36.7	39.6
T/K	709	737	814	845

$10^3 a$	**406**	**396**	**351**	**337**
γ	**2.71**	**2.64**	**2.34**	**2.25**

(b) The heat of the reaction is $H' - H^*$, and this can be found from equation 7.15

$$\partial \ln a/\partial T = \partial \ln \gamma/\partial T = -(H' - H^*)/RT^2$$

There is a slight variation of $H' - H^*$ with temperature, but taking an average over the whole range gives $-(H' - H^*) = -6.8 \text{ kJ mol}^{-1}$.

$$\therefore H' - H^* \approx \mathbf{7\ kJ\ mol^{-1}}$$

A8.9 Assuming a gas saturated with water vapour forms an ideal mixture, the chemical potential of water must be the same in the liquid and gaseous water.

$$\therefore \mu = \mu_L^* = \mu_G^* + RT \ln y_{sat}$$

which gives $\partial \ln y_{sat}/\partial T = \Delta_{vap}H/RT^2$ on differentiation, and hence

$$\ln y_{sat} = -\frac{\Delta_{vap}H}{R}\left(\frac{1}{T} - \frac{1}{T_0}\right)$$

$$= (-41\,900/8.314)(1/313 - 1/373) = -2.590$$

$$\therefore y_{sat} = \mathbf{0.075}$$

$p/P^{\ominus} = 55.13/760 = 0.0725$, in good agreement.

A8.10 From Raoult's law, $x = p/p^*$. In this case, $p = 1$ atm and p^* equals the vapour pressure of the liquefied gas at 273 K. p^* is found from

$$\ln \frac{p^*}{P^{\ominus}} = -\frac{\Delta_{vap}H}{R}\left(\frac{1}{T} - \frac{1}{T_0}\right)$$

$$\ln (p^*/P^{\ominus}) = (-\Delta_{vap}H/R)/(1/273 - 1/189) = 3.270$$

$$\therefore p^*/\text{atm} = 26.3 \text{ and } x_{sat} = P^{\ominus}/p^* = \mathbf{0.038}$$

(*N.B.* $P^{\ominus} = 1$ atm in this example since the 'normal' b.p. is used. Actual solubilities vary widely, *e.g.* water 0.00139, dimethylaniline 0.037.)

Chapter 9

Behaviour of Liquid Mixtures

9.1 MISCIBILITY

For ideal solutions $\Delta_{mix}G = \Sigma\, xRT \ln x$ (see equations 5.26 and 5.27).

$$\Delta_{mix}S = -\partial\Delta_{mix}G/\partial T = -\Sigma\, xR \ln x \qquad (9.1)$$

For mixing in a binary regular solution (equation 7.13)

$$\Delta G = B(x_1x_2^2 + x_2x_1^2) + \Sigma\, xRT \ln x$$

$$= Bx_1x_2 + RT(x_1 \ln x_1 + x_2 \ln x_2) \qquad (9.2)$$

$$\therefore \qquad \partial\Delta G/\partial x_1 = B(x_2 - x_1) - RT \ln(x_2/x_1) \qquad (9.3)$$

But since $\ln(x_2/x_1) \rightarrow -\infty$ as $x_1 \rightarrow 0$, $\partial\Delta G/\partial x_1$, and therefore $\Delta_{mix}G$, will be negative initially, whatever the value of B. This is true for other models as well and implies that, from a thermodynamic point of view, all substances must be miscible to some extent. (This would also be expected from general considerations of the entropy of mixing.) If $\Delta_{mix}H$ is negative or zero, miscibility will extend over the whole range of x, but if it is large and positive, ΔG may develop minima, resulting in a *miscibility gap*, a region of x in which the solution will split into two immiscible phases (Figure 9.1). This, of course, can never happen in the case of ideal solutions, since they have $\Delta_{mix}H = 0$ (equation 5.25).

The composition of the two immiscible phases will be determined by the requirement that, being in equilibrium, each component must have the same chemical potential throughout all phases in equilibrium (see equation 5.15), and, if referred to the same basis, the same activity. The joint tangent to any minima formed in the ΔG *vs.* x plot (Figure 9.1) will have intercepts

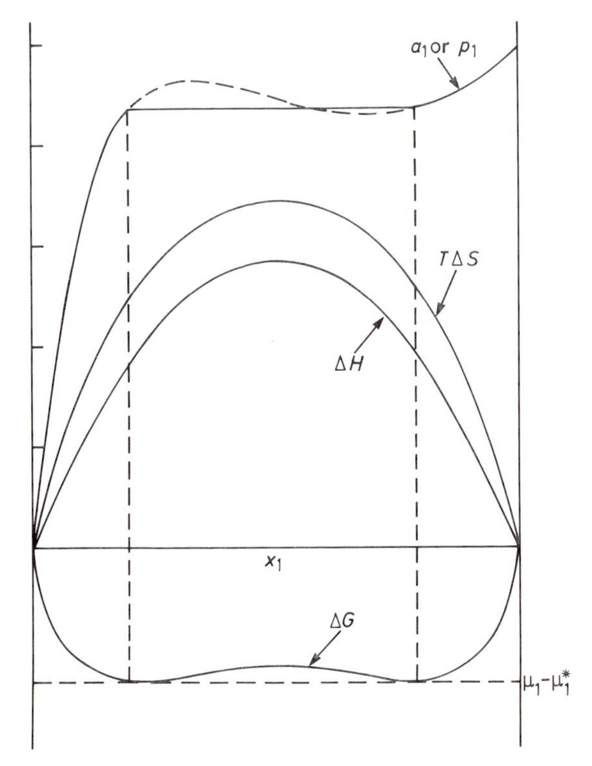

Figure 9.1a *Conjugate mixtures. The calculated v.p. curve, p_B, exhibits a maximum and a minimum, but the dotted part of the curve is not realized. The liquid splits into two immiscible phases at compositions at which the common tangent to the $\Delta_{mix}G$ curve touches. The intercepts of the tangent at x = 0 and 1 give $\mu - \mu^* = RT \ln a$ for the two components. Figure 9.1a is plotted for a 'regular' solution. In the general case, the tangent will not be horizontal, as shown in Figure 9.1b*

$\mu - \mu^* = RT \ln a$ (*cf.* $\Delta_{mix}H$), and therefore marks the composition of the immiscible phases, known as *conjugate mixtures*. It is important to realize that these conjugate mixtures not only contain components with the same μ and a, but also v.p., *etc.* Thus, although the solubility of benzene in water is minute ($x \approx 10^{-4}$), the v.p. of benzene in it is the same as that in the 'wet' benzene phase ($x \approx 0.9994$) in equilibrium with it.

With increasing temperature the $T\Delta_{mix}S$ term in the expression $\Delta G = \Delta H - T\Delta S$ becomes dominant (Figure 9.1), the miscibility

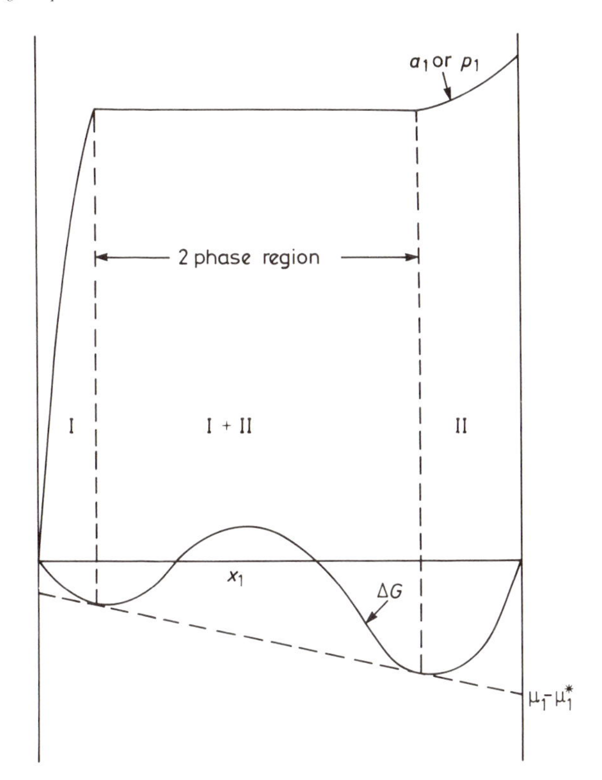

Figure 9.1b

gap decreases, and the minima merge at the upper *consolute* (or *critical miscibility*) temperature. Some systems exhibit a lower consolute temperature; this is rare and is due to the formation of a labile compound.

9.2 DISTILLATION

The total pressure P in a vapour pressure–composition diagram is obtained by summing the partial pressures. For an ideal system this is a straight line in a P *vs.* x plot, as shown in Figure 9.2. If x and y are the compositions in the liquid and vapour, respectively:

$$p_1 = x_1 p_1^* = y_1 P \text{ and } p_2 = x_2 p_2^* = y_2 P \tag{9.4}$$

$$\therefore \quad y_1 = x_1 p_1^*/P \text{ and } y_2 = x_2 p_2^*/P \tag{9.5}$$

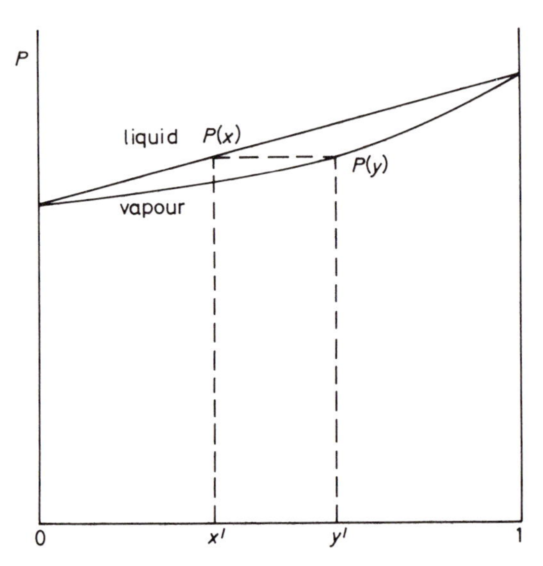

Figure 9.2 *Vapour–liquid equilibrium for an ideal solution (constant T). The partial pressures are linear with x, the liquid composition (Raoult's law), and the total pressure, P, is the sum of the partial pressures. The vapour composition, y, in equilibrium with the solution, x, (see equation 9.5) is connected with x by a horizontal 'tie-line'*

so that $y_2 > x_2$ if $P < p^*$, as seen in Figure 9.2, which also gives the total pressure as a function of the vapour composition, $P(y)$. The horizontal tie-line connects the liquid and vapour compositions in equilibrium. The *relative volatility*, α, is defined by

$$y_1/y_2 = \alpha x_1/x_2 \ (= x_1 p_1^*/x_2 p_2^*, \text{ if ideal}) \tag{9.6}$$

In non-ideal systems, vapours form almost ideal mixtures at low pressures, but the partial pressures of the liquid are now approximately given by

$$p_1 = a_1 p_1^* = y_1 P \text{ and } p_2 = a_2 p_2^* = y_2 P \tag{9.7}$$

and the P *vs.* x line will now be convex downwards for negative deviation and upwards for positive deviation from Raoult's law. In some cases this will lead to the formation of a minimum or maximum. The P *vs.* y line giving the vapour in equilibrium for each liquid composition can then be worked out from equation 9.7, given a or γ. Qualitatively it is given by *Konovalov's rule* that 'the vapour is always richer [than the liquid] in that component

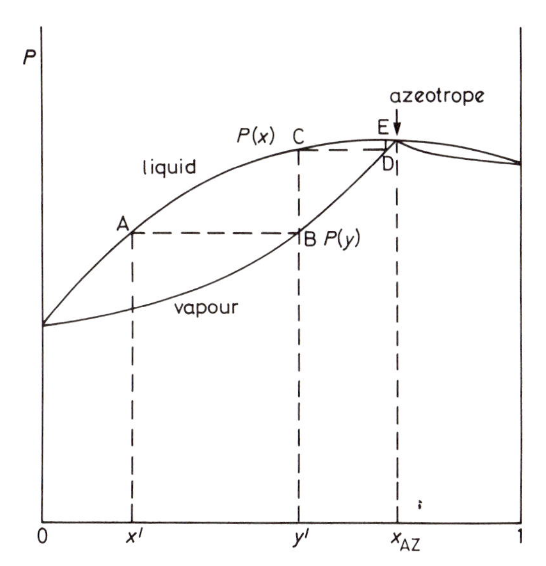

Figure 9.3 *Azeotrope formation. In a system with a large positive deviation from Raoult's law, the pressure–composition curve exhibits a maximum, known as the azeotrope. On distillation, a liquid A would give a vapour B which condenses to C,* etc., *and will eventually converge on the azeotropic composition. No further separation is then possible*

which, when added to the solution, increases its vapour pressure'.

9.3 AZEOTROPES

At a maximum or minimum in the P *vs.* x curve, the vapour is richer in neither component, *i.e.* the mixture distils unchanged. Such mixtures are known as azeotropes. They have the limiting composition that can be achieved by distillation of any mixture of intermediate composition (see Figure 9.3).

By rewriting equation 9.7 with $a = \gamma x$ and remembering that at the azeotrope $x = y$ it follows that

$$(\gamma_1)_{\mathrm{Az}} = P/p_1^*,\ (\gamma_2)_{\mathrm{Az}} = P/p_2^*,\ \text{and}\ (\gamma_1/\gamma_2)_{\mathrm{Az}} = p_2^*/p_1^* \quad (9.8)$$

These equations are useful in the determination of the van Laar on Margules constants (see equation 7.12).

If $p_1^* = p_2^*$, then $x = y$ over the whole range in ideal solutions, and azeotropes are inevitable when the two components have

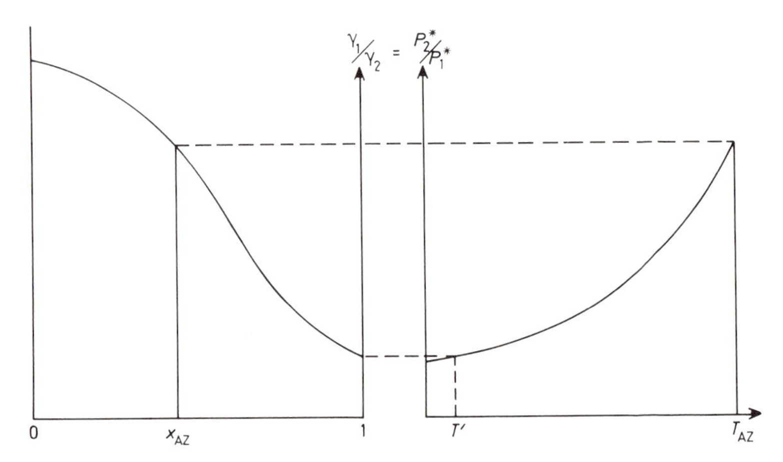

Figure 9.4 *Variation of azeotrope composition with temperature. A change in the temperature of distillation will alter the* p_2^*/p_1^* *ratio and therefore, by equation 9.8, the* γ_1/γ_2 *ratio at the azeotrope. If* γ_1/γ_2 *is plotted vs. x, and* p_2^*/p_1^* *vs. T (or 1/T) the T′ at which the azeotrope disappears at x = 1 can be read off from the appropriate horizontal tie-line*

nearly equal v.p. in non-ideal solutions. Resolution of azeotropes is therefore usually based on the principle of increasing the v.p. disparity between the two components by changing the distillation temperature and pressure, or by adding a third component which has a marked differential effect on the activities of the two original components.

The effect of temperature and pressure of distillation can best be demonstrated by reference to Figure 9.4. By plotting the ratio of γ values *vs.* x and the inverse ratio of p^* values *vs.* T on the same scale, tie-lines between these two plots will indicate the temperature at which the azeotrope will have disappeared (see equation 9.8).† The right-hand plot is obtained from the Clausius–Clapeyron equation (equation 4.4), with p replaced by p_2^*/p_1^* and $\Delta_{\text{vap}}H$ by $(\Delta_{\text{vap}}H_2 - \Delta_{\text{vap}}H_1)$.

Applied to the ethanol–water system, which has an azeotrope at 78.13 °C with 95.57 wt% ethanol, such a plot shows that at 33 °C the azeotrope will have moved out to 100% ethanol. This, of course would require distillation at low pressure (about 100 mmHg). (For other methods see reference 8.)

†It is assumed that the temperature variation of the γ_1/γ_2 ratio may be neglected.

EXAMPLES

9.1 In a solution obeying the equation

$$RT \ln \gamma_1 = Bx_2^2$$

what value of B will cause it to exhibit the phenomenon of 'critical mixing'?
(Use the fact that at critical mixing the solution is on the point of splitting into two phases in equilibrium.)

9.2 If the vapour pressure of a volatile component (mole fraction x) can be represented by the relationship $p = ax + cx^3$, show what limitations are imposed on the sign and magnitude of a and c if
(a) the relationship is to apply over the whole range of x and
(b) the system shows positive deviation from Raoult's law.

Show also that
(c) such a solution does not satisfy the condition for critical mixing, but may do so if a further term bx^2 is included in the above relationship for p.

9.3 Benzene and water have a limited solubility in each other. What type of activity/mole fraction plot would you expect from such a system?

If the two immiscible benzene–water phases contain 99.9 and 0.19 mol% water, respectively, at 20 °C, and water has a vapour pressure of 2.4 kPa at this temperature, what would be the activity coefficients of water in these two phases?

Estimate the vapour pressure of water over benzene containing 0.02 mol% water at this temperature.

9.4 Plot, on the same diagram, the curves for the activity a_1 and $\Delta_{mix}G/RT$ for a solution obeying the equation

$$RT \ln \gamma_1 = Bx_2^2 \text{ with } B/RT = 2.3$$

What part of the activity curve will be observable experimentally?

9.5 The aqueous layer in the water (W) – ethyl acetate (EtAc) system at 311 K contains 1.45 mol% EtAc. Making suitable assumptions (which should be stated):
(a) estimate the activity of W in the EtAc-rich layer, and
(b) if the total pressure over the immiscible layers is

26.66 kPa, estimate the activity of EtAc in either layer and the activity coefficient of EtAc in the water-rich layer.

$$p^*_{\mathrm{W}} = 6.40\ \mathrm{kPa},\ p^*_{\mathrm{EtAc}} = 22.13\ \mathrm{kPa}$$

9.6 The vapour pressure of pure n-heptane (H) at 298 K is 45 torr. At this temperature, H forms two immiscible layers with aniline which contain 5.5 and 92.6 mol% H, respectively.

Estimate the vapour pressure of H in mixtures containing respectively 3, 30, and 97 mol% H, stating your assumptions.

9.7 If the vapour pressure of one component of a binary mixture can be expressed in the form $p_1 = p^*_1 x^k_1$ and the vapour can be treated as an ideal gas, show that
(a) $p_2 = p^*_2 x^k_2$, and
(b) $y_1/y_2 = p^*_1 x^k_1 / p^*_2 x^k_2$
where y is the mole fraction in the vapour phase and k is a constant.

Calculate the value of k for an azeotropic mixture for which

$$x_1 = y_1 = 0.539,\ p^*_1 = 632\ \mathrm{torr},\ \text{and}\ p^*_2 = 583\ \mathrm{torr}$$

and deduce whether the system will show positive or negative deviation from Raoult's law.

9.8 The partial pressures of isopropanol (1) in mixtures with benzene and their total vapour pressures are given below for 298 K. Use these data to find the activity coefficients of the two constituents and estimate the composition of the azeotrope at 298 K.

$10^3 x_1$	0	59	146	362	521	700	924	1000
p_1/kPa	0	1.72	2.99	3.68	4.07	4.85	5.63	5.87
P/kPa	12.59	13.9	14.53	14.45	14.11	13.31	8.85	5.87

9.9 Calculate the activity coefficients at 70 °C from the data given below for that temperature for toluene (T) – acetic acid (A) mixtures.

$10^3 x_{\mathrm{T}}$	0	402	591	662	760	829	906	1000
p_{T}/kPa	0	15.71	19.36	20.76	22.30	23.49	24.81	26.93
p_{A}/kPa	18.13	12.76	10.43	9.24	7.71	6.20	4.07	0

$$\Delta_{\mathrm{vap}}H_{\mathrm{T}}/\mathrm{kJ\,mol^{-1}} = 33.5,\ \Delta_{\mathrm{vap}}H_{\mathrm{A}}/\mathrm{kJ\,mol^{-1}} = 23.7$$

Estimate the distillation temperature required to shift the azeotrope to 90.6 mol% toluene.

ANSWERS

A9.1 Phase equilibrium implies equal μ and a for each component. Critical mixing is the state just before the two phases develop, and the a *vs.* x curve (or the $\ln a$ *vs.* x curve) would have a horizontal inflexion about to develop into a separate maximum and minimum.

But $\ln a_1 = \ln x_1 + Bx_2^2/RT$

$$\therefore \partial \ln a_1/\partial x_1 = 1/x_1 - 2Bx_1/RT = 0$$

or

$$x_1^2 - x_1 - RT/2B = 0$$

for a horizontal tangent, and the quadratic will have a single root if $\boldsymbol{B = 2RT}$.

(Note that the solution is symmetrical in components 1 and 2, *i.e.* $x = 0.5$.)

A9.2 (a) For $x = 0$, $p = 0$, condition satisfied

For $x = 1$, $p = p^* \therefore \boldsymbol{a + c = p^*}$

(b) $p > xp^* \therefore ax + cx^3 > ax + cx$ or $c(x^2 - 1) > 0$ so that **c must be negative** ($x^2 \leqslant 1$)

(c) $\partial^2 p/(\partial x_1)^2 = 6cx = 0$ for critical mixing, but this is not so, and the equation cannot satisfy this condition without an extra term, which could be bx^2, making the second differential $2b + 6cx = 0$ with a single root for $x = -b/3c$.

A9.3 See Figure 9.1 for activity *vs.* x plots of immiscible phases.

For the water-rich phase the mole fraction of water $x = 0.999$. Assuming Raoult's law at high concentrations, $\boldsymbol{a = x = 0.999}$. In any case, a must be between 0.999 and 1.0. But the two phases are in equilibrium and therefore have the same activity, $a = \gamma x$ (Figure 9.1).

$$\gamma \approx \mathbf{1} \text{ at } x = 0.999,\ \gamma \approx 0.999/0.0019 \approx \mathbf{530} \text{ at } 0.19 \text{ mol\%}$$

At 0.02 mol%, γ will be about the same or slightly larger (positive deviation), so that

$$p = \gamma x p^* \approx 530 \times 0.0002 \times 2.4 = \mathbf{0.25\ kPa}$$

(This is equivalent to assuming a straight line to 0, *viz.* Henry's law, and could simply be $p = 0.999p^* \times 0.02/0.19$.)

A9.4 Since

$$\Delta_{\text{mix}} G = \Sigma\, xRT \ln x + Bx_1 x_2$$

(see equation 9.2), it is symmetrical in components 1 and 2 and need only be evaluated up to $x = 0.5$. The activities are also symmetrical about 0.5, but each individual component has its activity going from 0 to 1 as x goes from 0 to 1.

$$a = x \exp[(B/RT)(1 - x)^2] \text{ (see equation 7.13)}$$

The values are

$100x$	$10^3\Delta H/RT$	$10^3\Delta S/R$	$-10^4\Delta G/RT$	$10^4 a$	$100x$	$10^4 a$
5	109	199	893	3985	60	8669
10	207	325	1181	6443	65	8615
15	293	423	1295	7903	70	8610
20	368	500	1324	8719	75	8660
25	431	562	1311	9116	80	8771
30	483	611	1279	9259	85	8951
35	523	647	1242	9249	90	9209
40	552	673	1210	9155	95	9555
45	569	688	1189	9023	100	10 000
50	575	693	1181	8886		
55	569	688	1189	8763		
	⋮	⋮	⋮			

These figures are plotted in Figure 9.1 and show two minima in ΔG. The common tangent to these minima has intercepts $\mu - \mu^* = RT \ln a$, and therefore the same a for the two phases in equilibrium. It gives the concentrations at which two phases are formed, so that the activity between these two concentrations is constant and the curved dotted line is not realized in actual systems. It should be pointed out that the joint tangent to ΔG would not be horizontal in the general case. The curve pressure *vs.* x would be identical to the activity curve except that the vertical scale would be multiplied by p^*.

A9.5 It can usually be assumed that solutions at low concentrations obey Henry's law, and that they approach Raoult's law behaviour at high concentrations.

(a) Considering the aqueous layer of the mixture with 1.45 mol% EtAc and 98.55 mol% W, if Raoult's law applies to W, then

$a_W = x_W =$ **0.9855** and $p_W = 0.9855 \times 6.40 = 6.31$ kPa

(b) $p_{EtAc} = P - p_W = 26.66 - 6.31 = 20.35$ kPa

$a_{EtAc} = p/p^* = 20.35/22.13 =$ **0.92**

$\gamma_{EtAc} = a/x = 0.92/0.0145 =$ **63.4**

A9.6 With the data available it is only possible to approximate by assuming Raoult's law at high concentrations and Henry's law at low concentrations. Thus at 92.6 mol%, $a_H \approx x_H = 0.926$, and this figure would have to be the same for both phases, *i.e.* $a_H(x = 0.055) = 0.926$.

$p = p^*a$, so that $p_H = 45 \times 0.926 =$ **41.7 torr = 5.56 kPa** between 5.5 and 92.6 mol%. That is therefore the value at 30 mol%.

If there is a linear relationship between 0 and 5.5% (Henry's law)

$$p = 41.7 \times 3/5.5 \approx \mathbf{23\ torr = 3.1\ kPa}$$

Likewise, Raoult's law would give $45 \times 0.97 =$ **43.7 torr** at 97 mol%.

A9.7 (a) Since $a = p/p^*$ and using the Gibbs–Duhem equation, equation 5.18

$$\ln a_2 = -\int x_1/x_2 \,\mathrm{d}\ln a_1$$

with

$$\mathrm{d}\ln a_1 = \mathrm{d}(k \ln x_1) = k/x_1 \,\mathrm{d}x_1$$

$$\therefore \ln a_2 = -\int (x_1/x_2)(k/x_1)\,\mathrm{d}x_1 = \int k/x_2 \,\mathrm{d}x_2 = k \ln x_2 + I$$

However, the integration constant $I = 0$, since $a = 1$ for $x = 1$.

(b) $p = yP$, so that $y_1/y_2 = p_1^* x_1^k / p_2^* x_2^k$ QED

$$\therefore 539/461 = (632/583)(539/461)^k$$

or

$$(k - 1)\ln(539/461) = \ln(583/632) \therefore k = \mathbf{0.483}$$

With $a = x^{0.483}$, $a > x$ for all x, *i.e.* **positive** deviation.

A9.8 The activity coefficients are readily calculated from

$\gamma = p/xp^*$ with $p_1^* = 5.87$ kPa, $p_2^* = 12.59$ kPa, $p_2 = P - p_1$, and $x_2 = 1 - x_1$. The following results are obtained:

$10^3 x_1$	0	59	146	362	521	700	924	1000
γ_1	–	4.97	3.49	1.73	1.33	1.18	1.04	1.00
γ_2	1.00	1.03	1.07	1.34	1.66	2.24	3.36	–

The azeotrope position can be obtained either from the location of the maximum in the total pressure *vs.* x plot, or possibly more accurately from equation 9.8

$$p_1^*/p_2^* = (\gamma_2/\gamma_1)_{\mathrm{Az}} = 0.466$$

However, the data are too sparse to get an estimate better than **0.245** with confidence limits ±0.005 in x_1.

A9.9 For the first part refer to A9.8. The results are set out below.

$10^3 x_\mathrm{T}$	0	402	591	662	760	829	906	1000
p_T/kPa	0	15.71	19.36	20.76	22.30	23.49	24.81	26.93
γ_T	–	1.451	1.216	1.164	1.090	1.052	1.017	1
p_A/kPa	18.13	12.76	10.43	9.24	7.71	6.20	4.07	0
γ_A	1	1.177	1.407	1.508	1.772	2.000	2.388	–

The total pressure, $P = p_\mathrm{T} + p_\mathrm{A}$, clearly has a maximum somewhere between 0.662 and 0.760, but the exact position does not need to be found. It would have a γ ratio equal to the inverse p^* ratio, so that $(\gamma_\mathrm{T}/\gamma_\mathrm{A})_{\mathrm{Az}} = p_\mathrm{A}^*/p_\mathrm{T}^* = 0.673$ (see equation 9.8).

To estimate the temperature at which the azeotrope will have shifted to $x_\mathrm{T} = 0.906$ (for which $\gamma_\mathrm{T}/\gamma_\mathrm{A} = 0.426$), we make the usual assumption that the γ ratio is insensitive to small temperture changes. For this to become the azeotrope, the inverse p^* ratio will also have to be 0.426. The appropriate temperature can be found from equation 4.4. Subtracting the equation for p_T^* from that for p_A^* we get

$$\begin{aligned}\mathrm{d}\ln(p_\mathrm{A}^*/p_\mathrm{T}^*)/\mathrm{d}T &= (\Delta_{\mathrm{vap}}H_\mathrm{A} - \Delta_{\mathrm{vap}}H_\mathrm{T})/RT^2 \\ &= (-9800/8.314T^2)\mathrm{K}^{-1}\end{aligned}$$

The indefinite integral is

$$\ln(p_{\mathrm{A}}^{*}/p_{\mathrm{T}}^{*}) = 1179/T' + I \qquad (T' = T/\mathrm{K})$$

and the integration constant I can be found from the value at 70 °C:

$$\ln 0.673 = -0.396 = 1179/343 + I \therefore I = -3.833$$

$$\therefore \ln(p_{\mathrm{A}}^{*}/p_{\mathrm{T}}^{*}) = 1179/T' - 3.833$$

$$\ln 0.426 = -0.853 = 1179/T' - 3.833,\ T = \mathbf{396\ K}$$

Appendix A

Mathematical Requirements

EQUATION OF A STRAIGHT LINE

In an (x, z)-plane, any point on a general straight line must obey the equation $z = mx + c$, with $m = \tan \alpha = \Delta z/\Delta x$ as the slope and $c = z$ at $x = 0$ as the intercept (see Figure A.1).

If $z = \mathrm{f}(x)$ is any curve in the (x, z)-plane then the tangent at any point x' is a straight line $z = m'x + c'$ with a slope $m = \mathrm{d}z/\mathrm{d}x$ at x' and intercept c' (Figure A.2), or $\mathrm{d}z = m'\mathrm{d}x = (\mathrm{d}z/\mathrm{d}x)\,\mathrm{d}x$ for a small change.

PARTIAL DIFFERENTIALS

Most thermodynamic functions depend on several variables, say $z = \mathrm{f}(x, y)$. The slope of the tangent is easily determined if either x or y is kept constant [*i.e.* in a plane parallel to the (x, z) or (y, z)-planes respectively]; it is then described as a partial differential, and using the special symbol ∂ is written as $(\partial z/\partial x)_y$ or $(\partial z/\partial y)_x$, as appropriate. The subscript indicates which variable is kept constant. Thus, if $z = x^m y^n$, then the partial differentials of z are

$$(\partial z/\partial x)_y = mx^{m-1}y^n$$

and

$$(\partial z/\partial y)_x = nx^m y^{n-1}$$

It is easy to demonstrate in this case that differentiating the first result with respect to y or the second result w.r.t. x gives the same expression, namely $mnx^{m-1}y^{n-1}$, and it is a general rule for thermodynamic state functions that the result of cross-differentiation is independent of the order of differentiation, *i.e.*

$$\partial/\partial y(\partial z/\partial x)_y = \partial/\partial x(\partial z/\partial y)_x = \partial^2 z/\partial x \partial y \qquad \text{(A.1)}$$

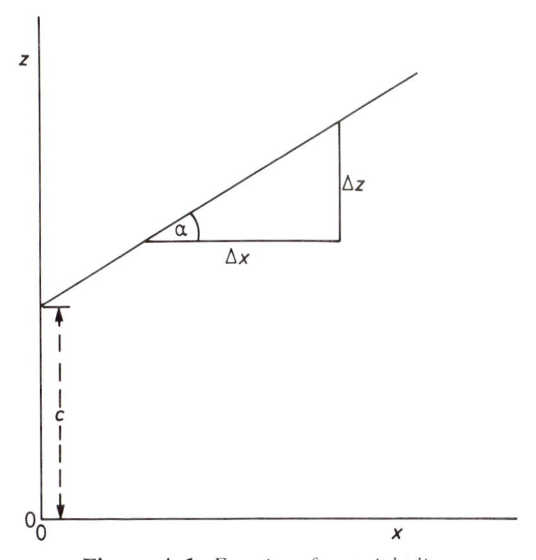

Figure A.1 *Equation of a straight line*

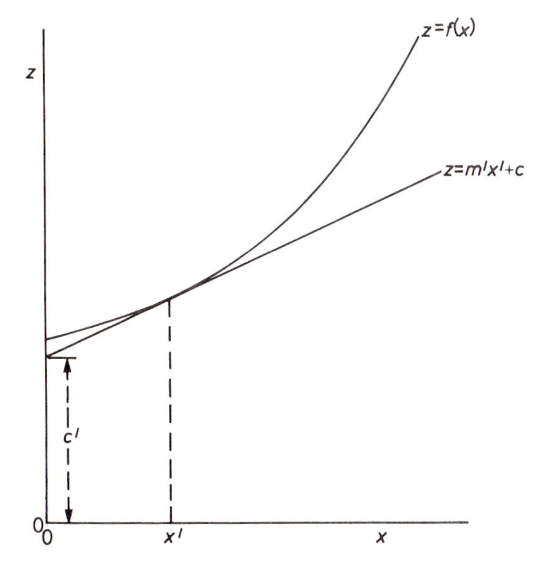

Figure A.2 *Equation of a tangent*

Example

Volume is a function of three variables: P, T, and n (the number of moles), *i.e.* $V = \mathrm{f}(P, T, n)$. The coefficients of expansion and compressibility are respectively

$$\alpha = V^{-1}(\partial V/\partial T)_{P,n}$$

and

$$\kappa = -V^{-1}(\partial V/\partial P)_{T,n}$$

Find them for an ideal gas for which $V = nRT/P$.

Answer

Note that such coefficients must be independent of the volume units and amount n (which has, however, to be kept constant). Treating P and n as constants, $\alpha = nR/PV$, but substituting $PV = nRT$ gives $\alpha = 1/T$, which is indeed independent of the amount.

Similarly, $\kappa = V^{-1}nRT/P^2$, and after substitution $\kappa = 1/P$.

INTEGRATION

It is important to think of integration not only as the inverse to differentiation, *e.g.*

$$\mathrm{d}(\ln x + c)/\mathrm{d}x = 1/x$$

$$\therefore \int 1/x\,\mathrm{d}x = \ln x + c \qquad \text{(A.2)}$$

but also as the summation of strips of area, *i.e.* in the example given, strips $1/x$ long and $\mathrm{d}x$ wide (Figure A.3). These strips will add up to the area under the curve, and in the case of a definite integral

$$\int_A^B 1/x\,\mathrm{d}x = [\ln x]_A^B = \ln B/A$$

add up to the area under the curve between A and B.

Note that all integrals have an implied integration constant, I, unless they have limits.

Since the order of addition does not affect the result ($a + b = b + a$) and integration is just an addition, a sum of integrals – with respect to the same variable – equals the integral of a sum, *e.g.*

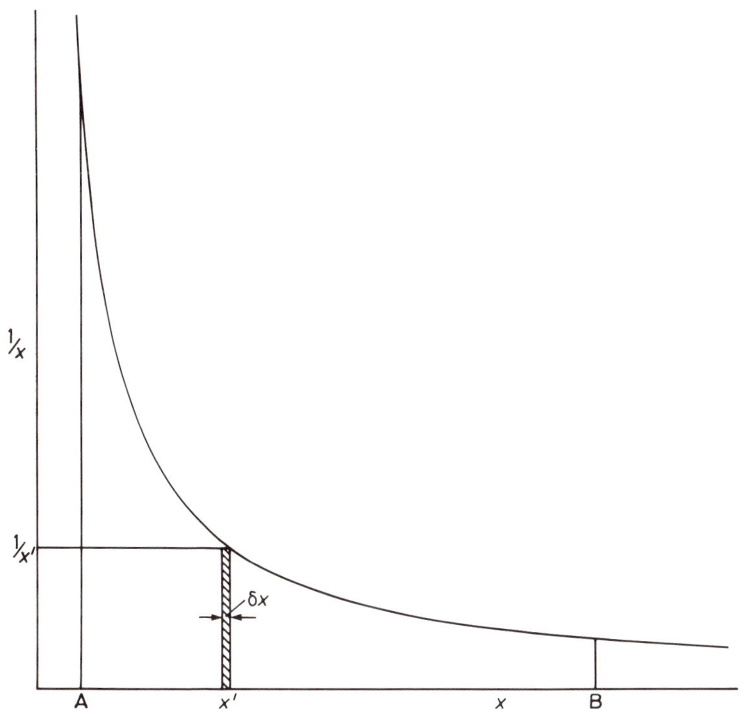

Figure A.3 *Integration*

$$\int f(x)\,dx + \int g(x)\,dx = \int [f(x) + g(x)]\,dx \qquad (A.3)$$

Similarly, for differential coefficients

$$d[f(x) + g(x)]/dx = d\,f(x)/dx + d\,g(x)/dx \qquad (A.4)$$

THE MATHEMATICS OF STATE FUNCTIONS

(It is assumed in this section that all functions considered are smooth and continuous. This is usually true for thermodynamic functions. A more detailed account may be found in reference 9.)

A state function is a function only of the state of the system, that is, only dependent on the specifications at the time and not on its history; any changes in it (going from one state to another) are therefore (by definition) independent of the path. The most convenient path is, of course, changing one variable at a time, so that

$$dz = (\partial z/\partial x)_y \, dx + (\partial z/\partial y)_x \, dy \qquad (A.5)$$

dz being called an *exact, total,* or *complete* differential.

The expression still holds for functions which *do* depend on the path, but, in that case, only for that *particular* path. To distinguish such *inexact* differentials, we use the symbol δ instead of d.

If dz is of the form $A\,dx + B\,dy$ then it can only be an exact differential (and therefore the differential of a state function) if it also satisfies equation A.5. Comparison shows that then

$$A = (\partial z/\partial x)_y$$

and

$$B = (\partial z/\partial y)_x \qquad (A.6)$$

and by equation A.1 the following relationship must therefore hold:

$$(\partial B/\partial x)_y = (\partial A/\partial y)_x \qquad (A.7)$$

Since any change in a state function is independent of the path (it depends on its initial and final states only), $\Delta z = z_{\text{final}} - z_{\text{initial}}$. The convention of writing it this way round conforms to the rules of integration, by which the change could also be followed along its entire path, still giving $z_{\text{final}} - z_{\text{initial}}$. This mathematical sign convention is always adopted in thermodynamics, *i.e.* Δ stands for final state − initial state. A drop in 'level', to which state functions are usefully compared, will therefore be represented by a *negative* change (*e.g.* a drop from 500 to 100 m will have Δ 'level' = −400 m).

If the final and initial states are the same, the change in the function is clearly zero, as in the case of a closed loop path. Formally, these propositions are given in equation A.8. They are not only true of all state functions, but the converse also holds, *i.e.* any function for which these propositions are true must be a state function.

$$\int_1^2 dz = z_2 - z_1 \text{ and } \oint dz = 0 \qquad (A.8)$$

The expression A.5 for dz may also be regarded as a shorthand way of writing any differential coefficient with respect to a chosen variable, *e.g.*

$$dz/dt = (\partial z/\partial x)_y \, dx/dt + (\partial z/\partial y)_x \, dy/dt$$

or, where t is not the only independent variable, differentiation with respect to t would become partial, so that

$$\partial z/\partial t = (\partial z/\partial x)\partial x/\partial t + (\partial z/\partial y)\partial y/\partial t \qquad \text{(A.9)}$$

The case frequently arises where $\mathrm{d}z$ in equation A.5 is zero, *i.e.* the total change in z is zero, while the partial changes are not, *e.g.* $V = \mathrm{f}(T, P)$, so that $\mathrm{d}V = (\partial V/\partial T)_P \mathrm{d}T + (\partial V/\partial P)_T \mathrm{d}P$ and this can be zero without $(\partial V/\partial T)_P$ or $(\partial V/\partial P)_T$ being zero. The two changes merely balance out, so that for constant total volume we have

$$(\partial V/\partial T)_P \mathrm{d}T + (\partial V/\partial P)_T \mathrm{d}P = 0 \text{ at constant } V$$

or

$$(\partial V/\partial T)_P(\partial T/\partial P)_V + (\partial V/\partial P)_T(\partial P/\partial P)_V = 0 \qquad \text{(A.10)}$$

Since $\partial P/\partial P = 1$ and $(\partial V/\partial P)_T(\partial P/\partial V)_T = 1$, equation A.10 can be rearranged to give the easily remembered 'triple product'

$$(\partial V/\partial T)_P(\partial T/\partial P)_V(\partial P/\partial V)_T = -1 \qquad \text{(A.11)}$$

so that the experimentally difficult to determine quantity $(\partial P/\partial T)_V$ can be obtained as the ratio of the readily available coefficients of volume expansion and compressibility $V^{-1}(\partial V/\partial T)_P$ and $-V^{-1}(\partial V/\partial P)_T$, respectively.

The general result, where $z = \mathrm{f}(x, y)$ and $\mathrm{d}z = 0$ in a particular case, gives

$$(\partial z/\partial x)_y(\partial x/\partial y)_z(\partial y/\partial z)_x = -1 \qquad \text{(A.12)}$$

Applications of the 'triple product rule' are given in the appropriate chapters.

Equation A.4 is equally true for sums and differences of functions of several variables; it is frequently applied to changes in state functions during a reaction, where the symbol Δ is used to denote the change from initial to final state. Thus, the volume change during a reaction would be $\Delta V = V_\mathrm{f} - V_\mathrm{i}$. For an ideal gas reaction at constant temperature and pressure involving a change in the number of moles of gas, $\Delta n = n_\mathrm{f} - n_\mathrm{i}$:

$$V_\mathrm{f} = n_\mathrm{f}RT/P \text{ and } \partial V_\mathrm{f}/\partial T = n_\mathrm{f}R/P$$

$$V_\mathrm{i} = n_\mathrm{i}RT/P \text{ and } \partial V_\mathrm{i}/\partial T = n_\mathrm{i}R/P$$

which on subtraction gives

$$\Delta V = \Delta nRT/P$$

and by equation A.4

$$\partial \Delta V / \partial T = \Delta n R / P \tag{A.13}$$

This last equation shows how the volume change for a constant temperature reaction behaves at different temperatures.

Appendix B

Constants and Units

FUNDAMENTAL CONSTANTS†

elementary charge	e	$1.602\,177\,33(49) \times 10^{-19}$ C
Planck constant	h	$6.626\,0755(40) \times 10^{-34}$ J s
Boltzmann constant	k	$1.380\,658(12) \times 10^{-23}$ J K^{-1}
Avogadro constant	L	$6.022\,1367(36) \times 10^{23}$ mol^{-1}
Faraday constant	$\mathcal{F}$	$9.648\,5309(29) \times 10^{4}$ C mol^{-1}
gas constant	R	$8.314\,510(70)$ J K^{-1} mol^{-1}
zero degrees Celsius	0 °C	273.15 K
standard pressure	$P^{\ominus}$	100 000 Pa
standard atmosphere	atm	101 325 Pa

SOME SI UNITS USED IN THERMODYNAMICS

Quantity	*SI Unit*	*Symbol*	*Dimensions in Base Units*
force	newton	N	m kg s^{-2}
pressure	pascal	Pa	m^{-1} kg s^{-2} (= N m^{-2})
energy	joule	J	m^{2} kg s^{-2} (= N m = Pa m^{3})
power	watt	W	m^{2} kg s^{-3} (= J s^{-1})

CONVERSION FACTORS

1 bar = 10^{5} Pa = 750 torr = 750 mmHg
1 atm = 760 mmHg = 1.013 25 bar = 101 325 Pa
1 mmHg = 1 torr = 133.322 Pa
1 cal = 4.184 J

†Figures in brackets indicate the uncertainty in the last two signicant figures given.

1 litre atm = 101.325 J
R = 1.987 cal K^{-1} = 0.082 058 l atm K^{-1} = 8.3145 J K^{-1} for 1 mol
1 kgf = 9.806 65 N
1 lb = 453.592 37 g
1 BTU = 252 cal = 1054.4 J

Appendix C

List of Symbols Used

(Page references follow most entries; IUPAC recommendations, where different, are given in parentheses.)

Symbol	Meaning
a	activity, 78
A	area, 2
	Helmholtz energy, 16
b	van der Waals constant, 66
B	virial coefficient, 66
c	concentration, 30
C_p	molar heat capacity at constant pressure, 3
C_V	molar heat capacity at constant volume, 3
$\mathcal{C}$	number of components, 58
d	relative density, 59
d	total differential, 1, 120
e	elementary charge, 123
$\mathcal{E}$	e.m.f., 28
f	fugacity (f, $\tilde{p}$), 67
F	force
$\mathcal{F}$	degrees of freedom, 58
	Faraday constant, 19, 123
g	standard gravity
g	gaseous, 9
G	Gibbs energy, 16
H	enthalpy, 3
I	integration constant, 8, 118
k	constant
K	equilibrium constant, 31
l	length
l	liquid (l, L), 9
	litre
L	Avogadro constant (L, N_A), 123
m	mass
	molality, 45
M	molar mass, 59
n	amount of substance, 4
p	partial pressure, 30
P	total pressure, 2
$\mathcal{P}$	number of phases, 58
q	heat absorbed, 1
R	gas constant, 4, 123
s	length of path
s	solid, 91
S	entropy, 16
t	Celsius temperature
T	absolute (thermodynamic) temperature, 1
U	internal energy, 1

v specific volume, 59
V total volume, 2
V' partial molar volume, 51
w work done by system ($-w$), 1
mass fraction, 59
x mole fraction, 43
X molar quantity (X, X_m), 3
y mole fraction of gas or vapour, 68
z charge number of an ion
Z compression factor, 66
α expansion coefficient, 118
general constant
γ activity coefficient (f)
fugacity coefficient (ϕ), 67
∂ partial differential operator, 3, 120
δ differential (non-state function), 1
Δ finite difference (final − initial), 3, 120
κ isothermal compressibility, 118
μ chemical potential, 56
μ_{JT} Joule–Thomson coefficient, 70
ν stoichiometric coefficient, 5, 31
Π product sign, 31
ρ density
Σ summation sign, 4

References

1 IUPAC, 'Quantities, Symbols and Units in Physical Chemistry', ed. I. Mills, Blackwell Scientific Publications, Oxford, 1988. (See also I. Mills, "What's in a name", *Chem. Br.*, 1988, **24**, 563.)

2 M. L. McGlashan, 'Physicochemical Quantities and Units', Royal Institute of Chemistry, London, 2nd edn., 1971.

3 C. W. Dannatt and H. J. T. Ellingham, *Discuss. Faraday Soc.*, 1948, No. 4.

4 J. A. V. Butler, 'Chemical Thermodynamics', Macmillan, London, 1951.

5 R. D. Freeman, *Bull. Chem. Thermodyn.*, 1982, **25**, 523.

6 J. H. Hildebrand and R. L. Scott, 'Regular Solutions', Prentice-Hall, New York, 1962.

7 O. A. Hougen and K. M. Watson, 'Chemical Process Principles: Vol. 2, Thermodynamics', 1st edn., Wiley, New York, 1947.

8 O. A. Hougen, K. M. Watson, and R. A. Ragatz, 'Chemical Process Principles: Vol. 2, Thermodynamics', 2nd edn., Wiley, New York, 1959, p. 922.

9 H. Margenau and G. M. Murphy, 'The Mathematics of Physics and Chemistry', Van Nostrand Reinhold, Princeton, New York, 2nd edn., 1964, chapter 1.

Bibliography

Some Compilations of Thermodynamic Data for $P^{\ominus}$

$P^{\ominus} = 0.1$ MPa:

D. D. Wagman *et al.*, 'The NBS Tables of Chemical Thermodynamic Properties', *J. Phys. Chem. Ref. Data*, 1982, **11**, Suppl. 2, 1.
M. W. Chase *et al.*, 'JANAF Thermochemical Tables', American Chemical Society for the National Bureau of Standards, New York, 3rd edn., 1986; also published as M. W. Chase *et al.*, *J. Phys. Chem. Ref. Data*, 1985, **14**, Suppl. 1.

$P^{\ominus} = 1$ atm:

J. B. Pedley *et al.*, 'Thermochemical Data of Organic Compounds', Chapman & Hall, London, 2nd edn., 1986.

Recent Textbooks

'Physical Chemistry':

R. A. Alberty, John Wiley & Sons, Chichester, 7th edn., 1987.
P. W. Atkins, Oxford University Press, Oxford, 3rd edn., 1986.
G. M. Barrow, McGraw-Hill, New York, 4th edn., 1979.
G. W. Castellan, Addison-Wesley, Reading (Mass.), 3rd edn., 1983.
G. N. Lewis and M. Randall, revised by K. S. Pitzer and L. Brewer, McGraw-Hill, New York, 2nd edn., 1961.

Other:

W. J. Moore, 'Basic Physical Chemistry', Prentice-Hall, New Jersey, 1983.
K. G. Denbigh, 'Principles of Chemical Equilibrium', Cambridge University Press, Cambridge, 4th edn., 1981.

M. L. McGlashan, 'Chemical Thermodynamics', Academic Press, London, 1979.
T. E. Daubert, 'Chemical Engineering Thermodynamics', McGraw-Hill, New York, 1985.

Index